# THE DEVELOPMENT OF ROTARY MOTION BY STEAM POWER

*Frontispiece: James Pickard's engine with a crank and flywheel researched and modelled by the author.*

# THE DEVELOPMENT OF ROTARY MOTION BY STEAM POWER

by
**DAVID K. HULSE**

© David K. Hulse 2001

Published by TEE Publishing Ltd., The Fosse,
Fosse Way, Leamington Spa CV31 1XN
Telephone 01926 614101

ISBN 1 85761 119 5

# ACKNOWLEDGEMENTS

With this book now complete the time had arrived to write the acknowledgement - a simple task I thought. How do I personally thank so many people who have helped me with this project, many of whom have now become friends.

The history of two engines is described in my second book, they are the worlds first engine to achieve rotary motion by the use of a crank and flywheel, and this is followed by a description of the worlds first engine to have its rotational speed controlled by a governor.

Hopefully all those persons who have helped me with these two engines will be mentioned in this acknowledgement.

The first of these engines to be researched and made in miniature was the engine shown on the cover of this book. The research started by contacting Michael Wright of the Science Museum at South Kensington who arranged for me to have unlimited access to the original engine preserved in the museum. He also supplied the two original drawings shown in the text dated 29$^{th}$ July 1788.

Also a big thank you must be given to the late Tom Walshaw and Dennis Chaddock these two professional engineers gave me practical advice which enabled an authentic model of the Boulton and Watt engine to be made.

Thank you to Ken Wood of the Stafford and District Model Engineering Society for the bronze medallion which is shown in Figure 84 and is to be permanently displayed with the engine.

David Yates of the Institute of Marine Engineers for allowing me to study all his research on Jonathan Hulls and this early proposal to power a sea going vessel with an atmospheric engine.

Steve Leadly a sculptor from Royal Doulton for modelling the bone china figures shown on each engine.

Paul Dyche a photographer who took the cover photographs which are used on both this and my first book called *The Early Development of the Steam Engine*.

Grateful thanks must be given to Norman Smedley and David York of TEE Publishing for their advice and editing of my original text, Norman read through my first draft and made many useful suggestions. A big thank you must also be given to Chris Deith for his decision to publish all my research, which will ultimately extend to three volumes.

And finally thank you to the person who has allowed me the time for all my thoughts and research notes to be arranged into this book. First reading through my initial text and making suggestions where it could be improved into a more readable form, my dear wife Julie.

David K Hulse

21 August 2001

# PREFACE

In David's first book '*The Early Development of the Steam Engine*' I commented upon the vast amount of research that David had undertaken to ensure that his superb models were as accurate as humanly possible.

Some years on after the first articles were serialised in '*Engineering in Miniature*' David continues to devote an immense amount of time to his research and has added further models to the original group. Many thousands of people have had an opportunity to view the models when they have appeared at various exhibitions and the interest they attract never seems to wane, nor indeed should it. The models, which are superb examples of the modeller's skill, are worthy of repeated examination and they are quite rightly to be described as museum quality models. David's dedication to detail continues to be evident with every component recreated exactly as on the prototype engine and the 151,000 bricks used thus far, all fired at different atmospheric conditions to ensure the very varied colouring seen in bricks of the period. Today, bricks can be fired very accurately giving a consistent colour but in the 18th century when some of these engine houses were built brick making itself was developing.

This latest book has shown us yet another side of David's commitment. Whilst the first book was produced in-house at TEE Publishing, this time David decided to go that extra mile by doing all his own typesetting. We are all most impressed with David's achievement because learning the intricacies of typesetting isn't easy for a newcomer, it's a specialist skill in its own way and hence our admiration. So, when you pick up this book to read, its worth remembering that David has not only researched and built the models, written the articles and edited the book but he's gone to the ultimate limit so that this really is, in the truest sense, David's own work - HIS book.

Incidentally, as I said at the outset, David's research is ongoing and there is, I believe, even more to be published. This book and its predecessor throw new light on the industrial revolution and those early engines. We look forward to working with David for many years to come to uncover the whole picture.

Chris Deith<br>
*Managing Editor*<br>
*Engineering in Miniature*

# CONTENTS

1 **How it all began.**
Rotary motion from an atmospheric engine .................. 4

2 **Rotation by a crank and flywheel.**
Snow Hill Engine, Sun and Planet gear, John Farey ......... 13

3 **Constructional details.**
Sequence of operation, Haystack boiler, Powering cylinder,
Piston seal, Operating mechanisms ........................ 23

4 **The main structural parts.**
Main operating beam, Crank and flywheel, Connecting rod,
Waterwheel technology, Constructional details ............ 44

5 **The Boulton and Watt Lap Engine of 1788.**
Double acting cylinder, Horse power, Original drawings,
Waggon boiler, Boiler safety valves ...................... 55

6 **The formation of a vacuum.**
Water cooled condenser, Double acting cylinder,
Nozzle housing, Operating levers ......................... 74

7 **Straight line movement from a piston rod.**
Parallel motion, Main beam of the engine, Connecting rod .. 89

8 **Rotary motion sun and planet gears.**
Flywheel, Speed regulation ............................... 99

9 **How the engine was assembled.**
Wooden frame, Maintenance instructions,
How the engine worked, Sequence of operation ............. 109

10 **Operating instructions.**
Starting the engine, Power measurement,
Soho Foundry and Manufactory, Boulton coinage ............ 117

11 **James Watt and his life in retirement.**
The final years, Garret workshop, Tributes to Watt,
Standard Boulton and Watt engine ......................... 124

12 **Soho Manufactory pictorial.** ............................ 135

**Bibliography** .............................................. 139

**Index** .................................................... 140

# LIST OF ILLUSTRATIONS

*Cover:* The model of James Watt's Lap Engine.
*Frontispiece:* James Pickard's engine

**The Pickard and Wasborough engine of 1779 and 1780.** ( numbers 1-42)
1   The miniature brick making machine designed and built by the author. 2
2   Denis Papin (1647-1712) ................................................................ 4
3   Thomas Savery's method of producing rotary motion ..................... 5
4   A treadle-operated lathe ................................................................ 6
5   Jonathan Hulls' proposal ............................................................... 7
6   Rotary motion from an atmospheric engine ................................... 9
7   John Smeaton's engine of 1777 .................................................... 12
8   James Pickard's patent specification ............................................ 14
9   James Watt's patent ..................................................................... 16
10  Sun and Planet ............................................................................. 17
11  Matthew Wasborough's flywheel and James Pickard's crank ........ 17
12  A rotary atmospheric engine ........................................................ 21
13  James Pickard's engine in model form ......................................... 23
14  A wooden inspection cover on a haystack boiler .......................... 24
15  Fire door beneath the boiler ......................................................... 25
16  A sectional view of the haystack boiler ....................................... 26
17  Complete haystack boiler before assembly into model ................. 27
18  The boiler header tank ................................................................. 29
19  Stand pipes to check the water level in the boiler ....................... 30
20  Two safety valves to prevent damage to the boiler ..................... 31
21  Lever to control the flow of steam from the boiler ...................... 32
22  A contemporary drawing of the eighteenth century ..................... 32
23  John Wilkinson's water-powered mill .......................................... 34
24  Blank of steel for the cylinder ..................................................... 35
25  Cylinder of the miniature engine in the lathe .............................. 36
26  The finished cylinder on the model ............................................. 37
27  A piston of a Newcomen engine .................................................. 37
28  The top of the cylinder ................................................................ 38
29  A valve used to control the flow of steam ................................... 39
30  A wooden header tank used for all the cold water ...................... 40
31  A drainage tank for the water from the cylinder ......................... 41
32  Tumbling bob used to turn on and off the water spray ................ 42

## LIST OF ILLUSTRATIONS

33 Tumbling bob used to control the flow of steam into the cylinder ......... 43
34 The main beam of the whole engine ..................................................... 45
35 The tension chains on the arch head ..................................................... 46
36 Weights used to return the piston to the top of the cylinder ................. 47
37 Lift pump used to supply water to the engine ....................................... 48
38 The flywheel and crank - the world's first ............................................. 49
39 The big end bearing and the connecting rod ......................................... 51
40 The centre of the flywheel water wheel technology .............................. 52
41 James Pickard and Matthew Wasborough ( bone china) ...................... 54

**The Boulton and Watt Lap Engine of 1788.** ( numbers 42- 92 )
42 The engine as displayed in the Science Museum in London ................. 58
43 The author's model of this engine ......................................................... 60
44 A model of the engine after extensive research .................................... 61
45 An illustration of a Lancashire bow saw ................................................ 62
46 A drawing signed by Matthew Boulton dated 29th July 1788 ................. 64
47 Another drawing dated 29th July 1788 .................................................. 65
48 Schematic drawing of a waggon boiler ................................................. 66
49 Safety valves to the boiler .................................................................... 68
50 High pressure safety valve ................................................................... 69
51 Mechanism for feeding the boiler with water ....................................... 70
52 Steam delivery pipe from boiler ........................................................... 72
53 Vacuum safety valve ............................................................................ 73
54 Water level measurement ..................................................................... 74
55 Condensing water tank with the assembled parts ................................. 76
56 Primary and secondary water pumps ................................................... 77
57 John Smeaton's water powered boring mill ......................................... 81
58 A cross section of the powering cylinder ............................................. 82
59 The cylinder showing all the valve operating levers ............................ 83
60 A view showing how the steam was controlled into the condenser ........ 86
61 The engine in the Science Museum. The valve operating levers ............ 88
62 The parallel motion of the engine in miniature .................................... 92
63 A standard ten horse power Boulton and Watt engine ......................... 95
64 The little end bearing ........................................................................... 97
65 The sun and planet gears on the original engine ..................................100
66 Sun and planet gears on the miniature engine .................................... 101
67 The flywheel ....................................................................................... 103
68 The main bearing of the flywheel on the original engine ..................,. 102
69 Thomas Mead's governing device ...................................................... 102

## LIST OF ILLUSTRATIONS

| | | |
|---|---|---|
| 70 | The governor which was fitted to the Lap Engine in 1788 | 105 |
| 71 | The governor which was fitted to the model | 106 |
| 72 | A photograph of the model showing the governor and flywheel | 108 |
| 73 | The main structural wooden framework | 109 |
| 74 | A sectional view of the engine showing how the parts fit together | 111 |
| 75 | A schematic drawing showing how the engine operated | 116 |
| 76 | The Soho Foundry | 119 |
| 77 | Boulton pennies and cast iron | 121 |
| 78 | Boulton coins made from copper | 121 |
| 79 | The workshop on the model built at one sixteenth full size | 122 |
| 80 | The parallel motion of the engine at Papplewick pumping station | 124 |
| 81 | Heathfield, James Watt's home at Handsworth Heath | 125 |
| 82 | The Garrett Workshop of James Watt | 126 |
| 83 | Sir Francis Chantrey's marble statue of James Watt | 127 |
| 84 | A bronze medallion showing Watt and one of his engines | 131 |
| 85 | The Soho Foundry of Peel and Williams, Manchester | 132 |
| 86 | A pictorial drawing by the author of the Lap Engine in 1788 | 133 |
| 87 | A standard ten horse power engine | 134 |
| 88 | The Soho Manufactory, opened in 1762 | 135 |
| 89 | A photograph of the Soho Manufactory | 136 |
| 90 | A photograph taken in 1861 | 136 |
| 91 | The Soho Manufactory being demolished in 1861 | 137 |
| 92 | James Watt's workroom at Heathfield | 138 |

# INTRODUCTION

My first book entitled the *Early Development of the Steam Engine* describes the mine pumping engines which were built between the years 1712- 1779, these two engines paved the way for the advancing industrial revolution throughout the world. They are Thomas Newcomen's Dudley Castle Engine - the world's first commercial steam engine and, James Watt's the Smethwick Engine - the first of Watt's three valve engines.

However, the two engines covered in this account have progressed from the pumping engines of the previous sixty-seven years into engines which could be used to provide the rotary power to industries which had previously been reliant on wind, water and horse power.

Manufacturers could now develop industries away from the restraints of natural sources, for example the textile industries could build factories away from the rivers which had been needed to drive the water powered mills.

The first of these two engines was an engine designed in 1779 by a Bristol engineer called Matthew Wasborough. This engine drove a manufactory in Birmingham which was owned by James Pickard. The manufactory was situated at the corner of Snow Hill and Water Street. When the Snow Hill engine was first installed it provided the rotary power to the machinery by the means of racks and pinions. However this method of obtaining rotary motion from a Newcomen type engine proved unreliable. In 1780 James Pickard removed this mechanism and fitted a simple crank and was so delighted with the result that he took out a patent for this method of operation on 23$^{rd}$ August 1780.

Matthew Wasborough at his own factory in Bristol had fitted a flywheel to one of his engines. When these two men combined their ideas and fitted a crank and flywheel on to the Birmingham engine, it became the first steam engine in the world to achieve rotary motion by this means.

The research into this engine was gathered together over a period of twenty years and the aim of this research was eventually to have enough information to enable a model of this engine to be made. With the research complete, the construction of the Pickard and Wasborough engine took approximately 2,800 hours to complete.

The second engine to be described is an engine which was built in 1788 to drive the lapping and polishing machines at the Boulton and Watt Soho Manufactory in Birmingham.

The engine became known by the task it performed from 1788 until 1858 and is now known as the Lap Engine. At the Soho Manufactory the engine drove forty-three lapping and polishing machines, and the operators were known as toy makers. The Lap Engine was designed by James Watt and was rated at ten horse power and was the first engine in the world to have its rotational speed regulated by a centrifugal governor. The original Lap Engine is now preserved in the Science Museum at South Kensington.

# INTRODUCTION

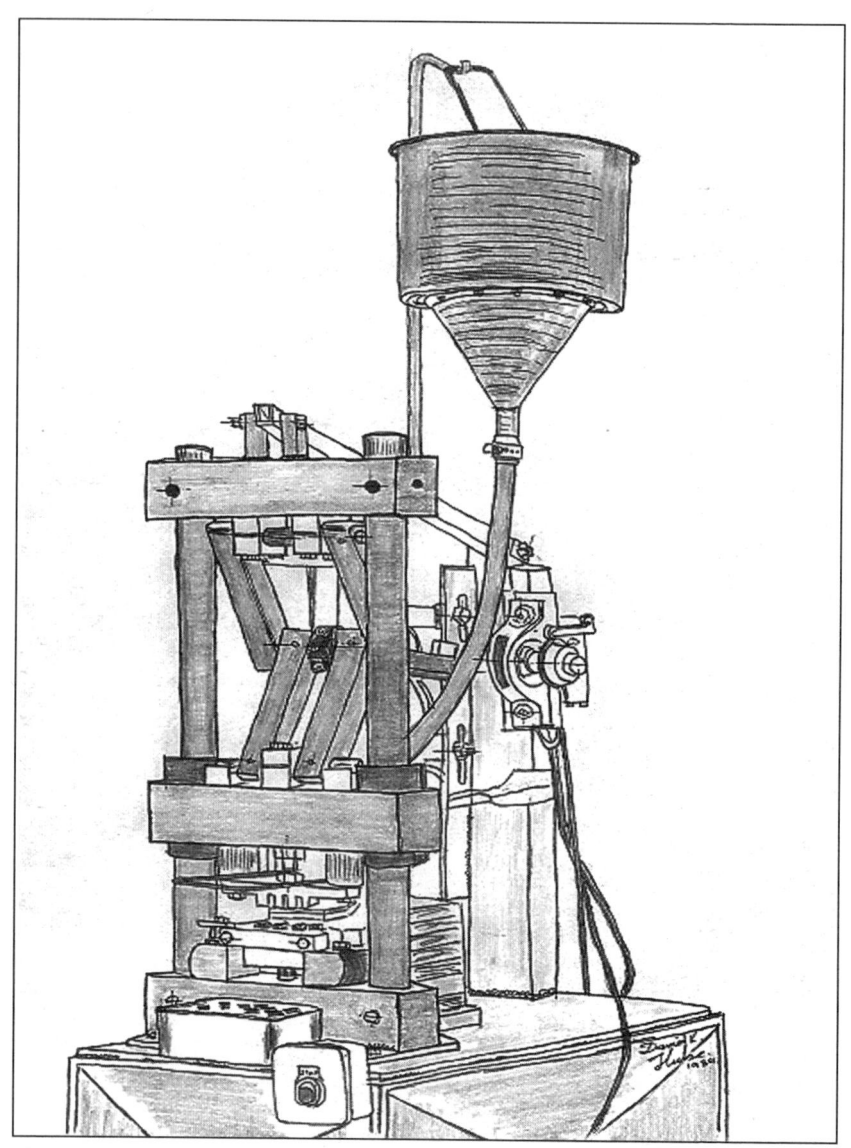

*Figure 1. A miniature brick making machine, designed and made to produce the ceramic bricks which have been used to make all the model engines.*

# INTRODUCTION

As with the engines researched and described in my first book the two engines described here have been made in miniature using the same materials as would have been available to the engineers of the eighteenth century. Castings have not been used; every component part is either fabricated or machined from the solid material. Also the miniature bricks were made on the machine shown as Figure 1. These two engines have been made to the scale of one sixteenth full size.

David K Hulse
16 August 2001

# CHAPTER 1
# HOW IT ALL BEGAN

**Rotary Motion by using the Pressure of the Atmosphere**
In 1690 Denis Papin made a proposal for an apparatus with two cylinders. The cylinders he proposed were to be fitted with free moving pistons which could be propelled within the bore of each cylinder by the creation of a vacuum on the underside of the piston. Papin's proposal was to have these pistons fitted with wrought iron rods into which had been formed gear teeth. These teeth could then engage with two opposing geared pinions.

On alternate strokes of the pistons a driven shaft would be compelled to revolve half a revolution and after both pistons had completed their powering strokes, one complete revolution would be made.

If his proposal could have been made to work continuously, rotary motion of the driven output shaft would have been the result. That Papin ever succeeded in applying his invention to any functional use on a large scale is unknown, because he was only a theoretical man, who had to rely upon other artisans to construct his proposed ideas.

In the 1790's the millwrights and engineers were not considered to have had enough experience to develop such a complex idea so different from their usual work. Papin goes on to say that in order to make cylinder bores accurate enough for his proposal to succeed, a specialised manufactory would have to be established for their production.

In a publication of 1709, Papin states that he did make such a machine in 1698, but only in model form: this model was accidentally destroyed before a full scale trial could be conducted.

*Figure 2. Denis Papin (1647 - 1712) Papin is possibly better known for his invention of the pressure cooker, which he first developed in 1679.*

## HOW IT ALL BEGAN

However, what it did show was that these early philosophers were now considering the steam engine as the provider of rotary motion. This was even before Thomas Newcomen had developed his first engine to drain the flood water from the coal mine in South Staffordshire.

In 1698 Thomas Savery, in an application for a patent, stated that his fire engine would be of great use for the 'working of all sorts of mills'. In the book which he called the *'Miner's Friend'*, published in 1702, he describes a plan for the working of mills with water wheels.

These wheels, he said, could be supplied with an artificial fall of water which had first been raised into a reservoir on higher ground by one of his fire engines. Savery's idea is shown in Figure 3.

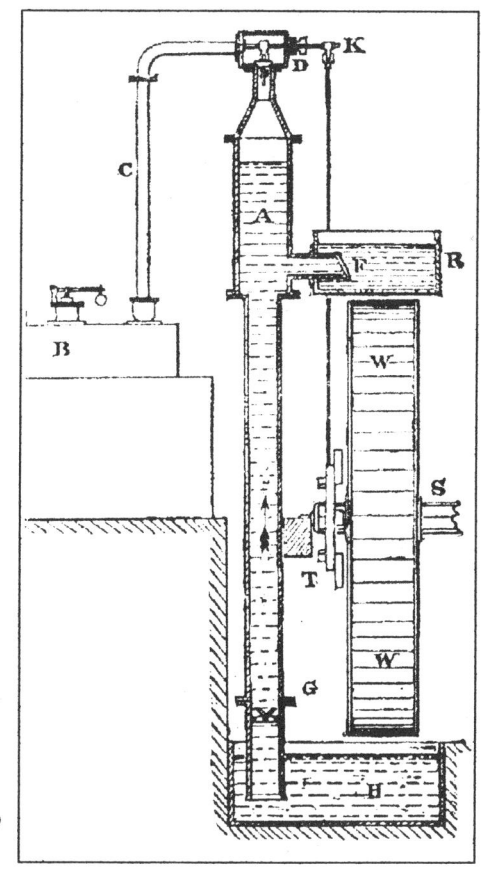

*Figure 3. Thomas Savery's proposal of first raising water high enough to flow over the top of an over shot waterwheel.*

After first rotating the wheel the water could then be lifted by one of his engines to the higher level, from where it could flow and be used all over again, and continuous rotary motion would be provided from this waterwheel to any process machinery for manufacturing.

Savery went on to say that direct rotary motion from a steam engine was a practical proposition, giving the example of the common foot-operated treadle lathe. This worked by converting the reciprocating motion of a foot treadle into rotary motion by the use of a lever or connecting rod and a simple crank. A flywheel was fitted and with this store of energy gave the lathe a uniform rotational speed. Figure 4, shows a foot-operated lathe of the early eighteenth century.

*Figure 4. A foot operated treadle lathe of the mid eighteenth century. This lathe is preserved at the Cheddleton Flint Mill in North Staffordshire*

However, these early philosophers could not visualise any way of applying a crank and a flywheel to a Newcomen engine which operated in such an unpredictable way.

In 1736, Jonathan Hulls obtained a patent from King George II for a machine to be used for towing ships out of, or into, harbours and rivers working against the wind or tide. He proposed to propel his vessel by rotating paddle-wheels. His proposal was to provide these paddle wheels with a continuous circular motion, and this circular motion was to be created by harnessing the powering force from a Newcomen type engine which had a cylinder open to the atmosphere. The straight line movement from the piston was to be transmitted onto each paddle wheel by passing a rope over two pulleys.

The pulleys were fixed on to an axle and the continuous rotary motion was

## HOW IT ALL BEGAN

*Figure 5. A patent was granted in 1736 to Jonathan Hulls for a machine for towing ships into and out of harbour. It is not known if this machine was ever made. However, this is the first recorded application where an atmospheric engine was to be used to propel a sea going vessel. This was drawn by Hulls in 1737 and is reproduced here by kind permission of the Institute of Marine Engineers.*

to be achieved by the returning stroke of the piston.

A counterweight was to be suspended on the other end of the rope, to return the piston to the top of the cylinder. It is not known if this plan was ever made, or any prototype was ever built. Figure 5 shows Hulls' proposal was on a grand scale, however, what the idea does show is that the straight line movement from a piston of an atmospheric engine was being considered to provide a constantly revolving output shaft.

In 1752, Mr. Chapman of Bristol established an extensive brass works where he used the rotary power from several overshot waterwheels. The water which was needed to drive these wheels flowed from a reservoir. After first turning the wheels, the water was returned to the reservoir by a large Newcomen engine.

This arrangement worked for about twenty years, and only ceased working on account of the large amount of coal the engine consumed. (A typical Newcomen engine with a working cylinder of 30 inches in diameter and a piston which made a stroke of 8 feet, would burn 24 cwt of coal in a twelve hour working day)

In the transactions of the Royal Society in 1759, Mr Kean Fitzgerald gave a proposal for a scheme to provide rotary motion to a ventilator. This ventilator was to provide a continuous supply of fresh air to the coal mines. His proposal was to use atmospheric engines, but he only considered this idea to be economical when the engines were also being used to draw off the flooded mine water. His proposed ventilator was a fan or vane continually revolving within a circular box.

This was also one of Denis Papin's proposals in 1690, which he called his *'Hessian Bellows'*.

A machine of this type required continuous circular motion in one direction and the object of Mr Fitzgerald's design was to obtain such a motion from the great oscillating lever of an atmospheric engine which operated at about twelve working strokes a minute. His plan was to make the fan rotate continually in one direction. The speed of the fan was to be increased to sixty revolutions per minute by some additional gearing.

He proposed to achieve his rotation by a combination of toothed racks engaging with two geared ratchet wheels. Each opposed to the other. These ratchet wheels were both mounted on a common axle. The racks were to be moved by an arc or toothed sector which was securely bolted on to the arch head of the main oscillating beam.

One of these ratchet wheels would revolve the axle for half a revolution, which was achieved on the powering stroke of the engine. During the return stroke of the piston to the top of the cylinder the second ratchet was compelled to rotate, but this time in the opposite direction. By driving through an arrangement of gears, the direction of rotation from this second ratchet wheel was reversed. Arranged in this way the up and down movement from the engines main oscillating beam did revolve the driven shaft one complete revolution. This provided continuous rotary motion in one direction from the single acting Newcomen type engine and this method is shown in Figure 6.

In 1762, a machine was made and constructed at the Hartley Colliery in Northumberland by Mr Oxley, for hauling coal out of the mine. This was an atmospheric engine which worked on a similar principle to Mr Fitzgerald's proposal of 1759. A sector with gear teeth was positioned at the end of the great lever which imparted reciprocating motion to a trundle wheel. The trundle wheel was fitted onto an axle which was driven by two geared pinions. These were fitted with ratchet-wheels and stops and, when arranged in this way, the driven axle revolved with a continuous circular motion. This arrangement used both the up and down strokes from the piston. The ratchets and stops were so arranged that the direction of rotation could be changed easily. This enabled the engine to wind up or lower down the corves (baskets) of coal within the mine shaft. This engine did not have a flywheel and is said to have revolved sluggishly and irregularly, with the result that the machinery was frequently damaged. The damage was caused by the violent motion to which this mechanism was subjected to on every complete stroke of the piston.

At a later date, the engine was modified to the tried and tested method of first pumping the water into a reservoir on higher ground. The water then flowed by

## HOW IT ALL BEGAN

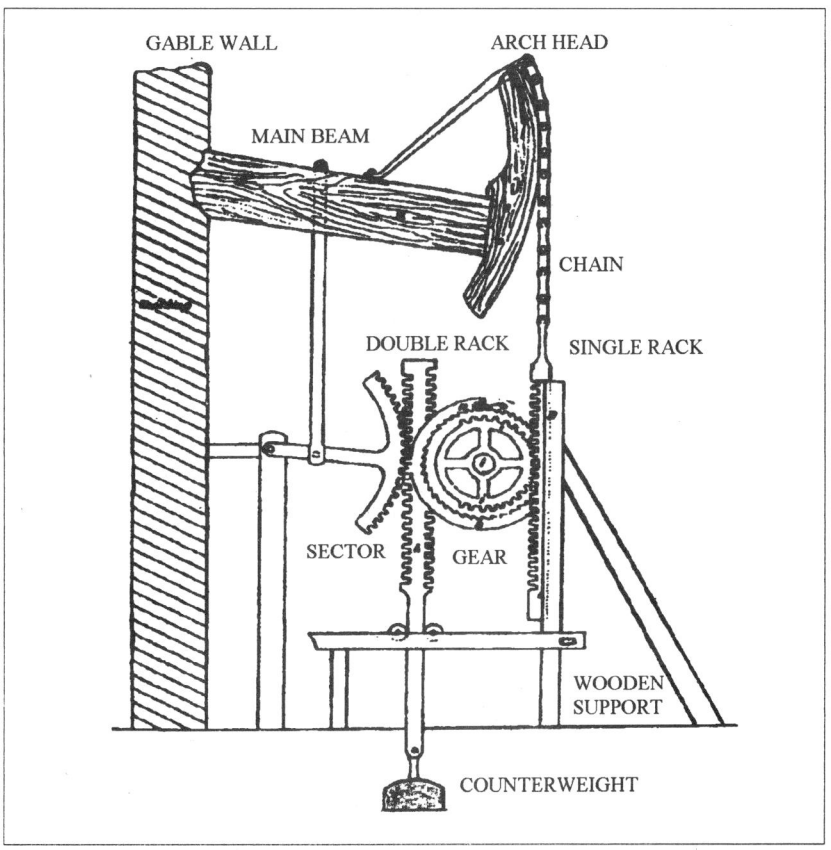

*Figure 6. This is the most widely known method which was used to obtain rotary motion directly from a single acting atmospheric engine in the eighteenth century.*

## HOW IT ALL BEGAN

gravity over the top of an overshot waterwheel, the rotary motion was then used to drive the factory machinery and, was also used to haul the coal out of the mine. If the Newcomen engine could have been made to reproduce the same length of a piston stroke on each operating cycle, this machine constructed by Mr Oxley would have been much more successful.

In 1769, Mr Dugald Clarke conceived an additional plan of obtaining continuous rotary motion from an atmospheric engine and he proposed that the plan should be applied to drive a sugar mill in Jamaica. Clarke's machine was first erected in London and a patent was granted for its method of operation.

This time the circular motion was produced by means of a toothed rack attached on to the great lever of his engine. The rack was compelled to ascend and then descend in a vertical groove.

This gave linear motion to two other racks fixed on the same axis, which then transferred the motion onto a horizontal cogwheel of the sugar mill by an arrangement of trundle wheels. As the toothed rack moved up and down the axis was alternately rotated by one of the geared wheels and, by the means of a catch, the other geared wheel was turned backwards by the teeth on the rack.

The difficulty with this plan was its complexity. It proved to be very unreliable because of the unpredictable nature of the Newcomen engine. (All these early attempts to harness the power of a Newcomen engine to produce rotary motion appeared to be variations on the rack and geared sector principle, but this arrangement by Clarke gave a very irregular and jerky motion).

In 1777 a paper was presented to the Royal Society in London by Mr John Stewart in which he described another plan for converting the reciprocating motion of an atmospheric engine into a continuous circular motion. This, he said, could be used to provide the rotary power to all types of mills. Stewart's idea was to have two endless chains circulating over pulleys. These pulleys were fitted onto a strong wooden frame. The chains were to move up and down by the oscillating motion of the main beam, in a similar manner to the domestic sash-window. The two chains were anchored on to the teeth at the opposite sides of the main gear wheel in such a manner as to give it a circular motion. First one chain and then the other was acting alternately on the opposite sides of the wheel. One chain impelled the gear wheel to revolve half a revolution during the powering stroke of the piston and the other chain completed the revolution when the piston was returning to the top of the cylinder.

John Stewart, who developed this idea, stated that if the main gear wheel just described was applied to the main axis of a flour-mill, the reciprocating motion from the engine would keep the mill continually revolving. Stewart states that the millstone was heavy and would serve as a flywheel and this would regulate any variations in speed, which might have occurred in the directional change at the start of each powering stroke of the piston. 'In sawmills, or any other mills which do not give a great velocity to some heavy rotating bodies which could serve as a regulating power, a large flywheel must be fitted'.

# HOW IT ALL BEGAN

In his paper he speaks of a crank or winch as a method of obtaining the circular motion which he says occurs naturally in theory, but in practice he thought this idea was impossible because of the irregular motion of the atmospheric engine. Stewart went on to say the engine could not be relied upon to make an accurate and repeatable length of a piston stroke through the day. Smeaton said the first time such an engine was fitted with a crank the whole machine would be broken to pieces.

Mr. Stewart's paper was referred by the Council of the Royal Society to Mr Smeaton, who remarked upon the difficulty arising 'from the absolute stop of the whole mass of moving parts, as often as the direction of the motion was changed. Though a flywheel might be applied to regulate the motion, it must be such a large one as would not be readily be controlled by the engine itself, and he considered that the use of such a flywheel would be a greater incumbrance to a mill, than a waterwheel to be supplied by water pumped up by the engine'.

**The year 1779 and the goal is almost in sight.**

The single acting atmospheric pumping engine, first developed by Thomas Newcomen in 1712, proved successful in removing the flood water from the coal and mineral mines and was used in this simple form for more than sixty years. However the engineers and philosophers of the eighteenth century were at a complete loss to imagine such an engine ever producing a reliable circular motion.

When an atmospheric engine was working, drawing the flood water from the mines, the resistance of the pump rods and also the amount of water removed on each stroke, governed the working speed of the engine. The engines working on the Newcomen principle did not even make the same length of a piston stroke throughout a working day.

The problem facing the engineers was how to make an engine produce a reliable movement so necessary in a rotary engine, when taking into account all these variables. Many atmospheric engines were adapted to provide indirect rotary motion. The most common method used was for an engine to recirculate water by pumping it high enough to flow by gravity over the top of an overshot waterwheel. The rotation of the wheel was then harnessed to provide the rotary power to drive each individual piece of process machinery. This method of providing rotary power was only considered economical if the engine was also used to remove the flood water from the mine workings.

Before flowing away to waste the water could also be used to turn the water wheel which was then connected to each individual machine at a factory.

After the initial cost of the installation the resultant rotary motion was an added bonus. However, when an engine had to be installed just to recirculate water, the rotary power was expensive because the Newcomen engine consumed vast amounts of coal.

However at this time the engineers thought it was the only way to harness the power of the atmospheric engine, taking into account all its idiosyncrasies. The resultant rotary motion was classified as an indirect method. What was proving difficult for engineers to imagine was how to harness the direct linear movement from the engine into one of rotary motion to drive the process machinery.

Even that great engineer John Smeaton stated that if reliable rotary motion was needed, the Newcomen engine must first have to pump water over the top of a waterwheel. Figure 7 shows a typical atmospheric pumping engine designed by John Smeaton in the late 1770's.

*Figure 7. A large atmospheric engine which was designed by John Smeaton in 1777. This installation is shown with a haystack boiler outside the main engine house.*

# CHAPTER 2

# ROTATION BY A CRANK AND FLYWHEEL

**James Pickard and Matthew Wasborough.**
In 1779, Matthew Wasborough, a Bristol engineer, designed a mechanism to harness the up and down movement created by a Newcomen engine to provide direct rotary motion. On an engine of his own construction, working on the Newcomen principles Wasborough fitted a ratchet and pawl arrangement to convert the up and down movement of the main oscillating beam into a rotary movement of a driven shaft. From this shaft, he was able to drive his machine tools, such as lathes, at his factory in Bristol.

Matthew Wasborough was then commissioned by a Birmingham manufacturer, James Pickard, to build a rotary engine based upon the design of his engine at Bristol.

The ratchet and pawl arrangement, whilst providing rotary power to Pickard's factory but proved very unreliable. The difficulty with this type of mechanism was keeping the mating teeth in an accurate mesh, as the two gears were moved apart by the great force needed to rotate the driven shaft. When this occurred the gears became disengaged and the rotary drive was lost.

Matthew Wasborough took out a patent on the 6th July 1779 for his own ideas on converting an engine to provide rotary motion. However, in his patent, he covers a device which he says would regulate and render the rotary motion more uniform. He called this device a *fly*. Wasborough stated that when his engines were used with this innovation to create rotary motion and, also fitted with his flies (flywheels), they could be adapted to drive a variety of process machinery such as grinding, bolting, spinning, drilling etc. Wasborough also stated that he intended to apply his engine for the purpose of moving ships, boats, or any vessel on the water. At the same time in Birmingham, James Pickard was contemplating a way of improving the reliability of his engine.

The ratchet and pawl arrangement was proving too difficult to maintain when attempting to keep his factory in constant production. He began to think of the result he would get if he removed the entire mechanism, which had been designed by Matthew Wasborough, and replaced it with a simple crank. Pickard went ahead with this idea and modified his engine at Snow Hill by the fitting of this simple mechanism. The result was so promising that he took out a patent for this idea on the 23rd August 1780. When this patent became law, it gave James Pickard the sole rights to its use on steam engines for the following twelve years. At this stage in its development his engine was thought to have been fitted with the two contra rotating gear wheels, which are shown in the patent specification of 1780 and reproduced in Figure 8 shown on page 14.

The smaller gear with the additional weight fitted would, hopefully, revolve with enough inertia to return the piston to the top of the cylinder from where another powering stroke could be made. It was not known how successful the engine was at this stage in its development. However, the engine was soon to be

# ROTATION BY A CRANK AND FLYWHEEL

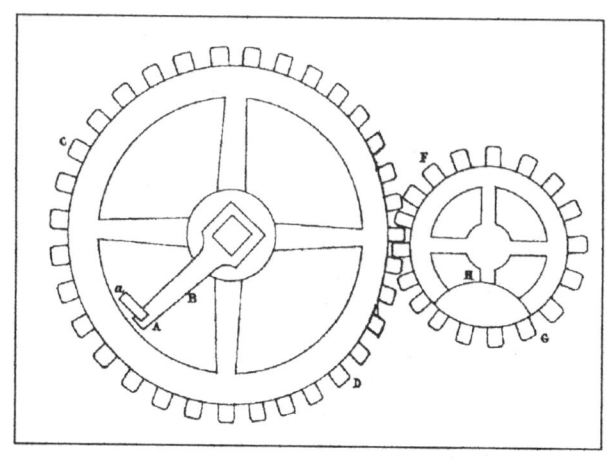

*Figure 8. A drawing taken from James Pickard's patent number 1263 showing a crank with two contra rotating geared wheels.*

a complete success because Matthew Wasborough allowed James Pickard the right to use his idea of the flywheel.

With these two elements fitted, the crank and the flywheel, the Newcomen engine at Snow Hill in Birmingham became the first engine in the world to produce rotary motion directly and reliably by this new method. The goal, so long sought after by the engineers of the previous seventy years, had finally been achieved and the unpredictable nature of the Newcomen engine was at last under control. The simple crank compelled the engine to make strokes of equal length and the flywheel regulated the fluctuating forces within one cycle of the engine. The flywheel also helped the piston to return to the top of the cylinder to enable another powering stroke to begin.

The granting of this patent to James Pickard on 23rd August 1780 caused a great controversy. Many said that Pickard was not the inventor of this method of achieving rotary motion by a steam engine. Matthew Boulton and James Watt said he should not have been granted a patent because the idea was stolen from their organisation by an angry workman.

**Boulton and Watt - the news is broken\***

James Watt does not appear to have been seriously perturbed by the news of this development. The reason was that he did not consider the application of the crank to the steam engine was a patentable invention.

Matthew Boulton was on business in Cornwall at this time and, Watt reported the news to him as follows: 'Matthew Wasborough has got a single crank to the Snow Hill engine and it does very well, better than the rick rack. I think you should call on Matthew as you return and let him know that we will dispute his exclusive right to these cranks'

\* *Taken From 'James Watt' by L.T.C. Rolt.*

## ROTATION BY A CRANK AND FLYWHEEL

Evidently, Wasborough was also in Bristol and, in the light of what was to follow, it must be assumed that when Watt wrote this letter he had not seen the Snow Hill engine for himself, but was passing on what he had been told. Boulton wrote in reply: 'I think the double cylinder and crank are better than any of them and, if one were erected it would crush these quacks and, it is a very desirable thing to do'. He went on to ask Watt to instruct their London solicitor to watch out for any patent application which Wasborough or Pickard might lodge so that they could oppose it if they had to.

This advice came too late. Boulton wrote on 26th November 1780 that a patent had been obtained in the name of James Pickard on 23rd August which became a lawful document on 9th December 1780.

The specification was a simple one and included a single diagram showing the crank, the two gear wheels and the counterweight. James Watt was furious and a bitter controversy ensued. Although, in the event, Boulton and Watt never contested Pickard's patent, they evidently contemplated doing so, for a document was prepared, presumably for counsel's opinion, which summarises the basis for Watt's allegations. This asserts that, during the summer of 1779, an experimental model was made at the Soho Manufactory which featured the crank, gearing and counterweight exactly as shown in Pickard's patented specification.

Their employee, Richard Cartwright, had often seen and worked this model and had communicated the principle of it to Samuel Evans, Pickard's engineman at Snow Hill. Watts' own words confirm that it was not the principle of the crank which was stolen from him, but the idea of the two-to-one gearing and the counterweight. This was an idea which Watt could well afford to discard. His new double-acting engine largely solved the problem which had inspired it and it was, in any case, a clumsy substitute for a simple flywheel. If, as it now appeared, James Watt thought so highly of it, why had he not patented the idea himself?

This was evidently recognised as a point of weakness in the case, for it was answered in the statement prepared for counsel. Preoccupation with engine building in Cornwall was the weak reason given, whereas it is much more likely that Watt considered the idea not worth patenting and had discarded it before the storm blew up. The gear and counterweight were very soon abandoned on the Snow Hill engine in favour of a flywheel in fact some writers express the opinion that it was never used at all.

The point which needs emphasis in this complex affair is that what Pickard had patented was the crank in combination with two gears, and onto the smaller gear he attached a balance weight. James Watt was almost certainly correct in his original belief that the application of the crank alone was not an original idea, and should not have been granted a patent for its application on a steam engine.

Had he held firmly to this belief he would have put the matter to the test by building an engine with a crank but this he now obstinately refused to do, insisting that the Pickard patent blocked such a course.

# ROTATION BY A CRANK AND FLYWHEEL

Pickard and Wasborough again asked for a licence to build the Watt engine and they not only offered to pay for this, but also declared that in return Boulton and Watt might freely use the crank on their machines. But Watt rejected this olive branch. He would be content with nothing less than a half share in the Pickard patent, a proposal which was naturally refused since its acceptance would have amounted to an admission that the idea had been stolen.

Having made up his mind there was no moving Watt and the crank was never used on any Boulton and Watt engine until the patent had expired. Watt set to work to design ways of avoiding the hateful patent and, in October 1781, he patented no less than five different ways of producing rotary motion without using a simple crank.

Three of these ideas need not be described - they were thinly disguised variants of the crank principle and, on this account they were never used.

The fourth was a swash plate arrangement of which there is a model in the Science Museum. This was fitted experimentally to one of the Soho engines but nothing more was heard of it which was probably just as well. Figure 9 shows a

*Figure 9. Swash plate James Watt's fourth method of obtaining rotary motion without infringing Pickard's patent.*

model of this idea which is thought to have been made at the Soho Manufactory.

Finally, there was the sun and planet gear which became the standard on all Watt rotative engines until 1792 when the Pickard patent finally expired. In this arrangement the planet wheel is fixed to the connecting rod and circles the sun wheel mounted on the drive shaft thus compelling the latter to rotate. The sun and planet gearing was first fitted experimentally to a small 15 inch diameter single-acting engine at the Soho Manufactory, which was used to drive a tilt hammer. When Watt had overcome the initial teething troubles, the first rotative beam engines were designed and made for other manufacturers and were then installed at manufactories throughout the country.

There was a much larger engine designed to drive a tilt hammer for Wilkinson's Bradley Forge (1783) and a small winding engine was installed

## ROTATION BY A CRANK AND FLYWHEEL

*Figure 10. The sun and planet method of obtaining rotary motion. This photograph was taken from the Lap Engine in miniature.*

*Figure 11. Matthew Wasborough's flywheel and James Pickard's crank fitted on to the model of the engine.*

at Wheal Virgin, a copper mine in Cornwall (June 1784). Engines were also supplied to two London brewers, Goodwyn and Whitbread, and another engine to drive a tilt hammer at the Horsehay Forge in Shropshire. These three engines were commissioned during the winter of 1784-1785. Figure 10 shows the sun and planet arrangement of obtaining rotary motion.

When originally built all these engines were single acting and in every case the connecting rod end of the beam was heavily weighted. This was to assist the action of the flywheel in returning the piston to the top of the cylinder to enable another powered stroke to commence.

Boulton suggested that a heavy cast iron connecting rod would help the engine to run more smoothly and moderate the chatter in the sun and planet gears. However Watt rejected this idea and wooden connecting rods continued to be the general rule on Boulton and Watt engines until c1800.

Shortly after the Bradley Forge engine was commissioned in the spring of 1783 Watt had the first double-acting rotative engine running at the Soho Manufactory. This was the engine previously referred to which had the rack and sector motion between the piston rod and beam. Watt always gave any new engine development a full-scale trial within the four walls of the Soho Manufactory and only tried the engine out for the customer after he had carried out any modifications which the trials might reveal. Although he was in many ways over cautious and too conservative this policy of Watt's deserves nothing but praise. It contributed greatly to the solid reputation which the Boulton and Watt organisation gained during the early development of the steam engine.

The first double-acting rotative beam engine to be built for a customer illustrates this point very well. When Watt designed an engine for Messrs Cotes & Jarratt of Hull in June 1784 the arrangement drawing featured the rack and sector mechanism in order to achieve a parallel action to the piston rod. However when the engine was actually delivered, the three-bar motion was substituted. (Watt's parallel motion had been fitted ). The engine was erected at the Soho Manufactory and perfected before being dismantled and reassembled at Hull.

Whoever, was the true father of the crank method of obtaining rotary motion from the reciprocating cycle of a steam engine, great credit must be given to James Pickard and Matthew Wasborough for having the courage to fit a crank on to an engine which worked in such an unpredictable way. Once fitted the crank was now controlling the operating cycle of the engine. The scholars and engineers who preceded Pickard and Wasborough had always thought that rotary motion provided by an atmospheric engine had to take into account the engine's unpredictable movements.

However, once fitted, the crank converted the up and down movement from the connection rod into the rotary motion of a driven output shaft (crankshaft) With the addition of Wasborough's innovation of a flywheel, the regularity of rotation from the engine at Snow Hill was greatly improved. These two men had converted the Newcomen engine, which had worked for the past sixty-eight years drawing the flood water from the coal and mineral mines into an engine which could be used to provide the rotary motion to all manufacturing industries. James

# ROTATION BY A CRANK AND FLYWHEEL

Watt never ceased to show his anger at James Pickard for using and patenting the crank. However, later in life Watt candidly admitted that he had not appreciated the great effect a flywheel would have on a rotating engine, and he did give credit to Matthew Wasborough for being the first engineer to fit one onto the crankshaft of a steam engine. Figure 11 shows the centre of the flywheel and the crank on the model engine.

## What is known of James Pickard and his Engine?

Not much is known about James Pickard and his factory which was situated at the corner of Snow Hill and Water Street in Birmingham. The factory was built some time before 1779 and must have produced a great variety of goods, ranging from very small domestic items to large components which could be used in the engineering industry. In his patent application for the crank James Pickard actually states his occupation as that of a button maker. In the Boulton and Watt papers now housed in the Birmingham Reference Library there is a manuscript written by John Southern who was a Boulton and Watt employee.

Southern records a visit he made to James Pickard's factory at Snow Hill in 1780. In this report Southern says the engine had a powering cylinder with a working diameter of 30 inches. He also states that the crank measured 3 feet 7 inches. If the engine was constructed to the usual proportions with the main operating beam pivoting centrally on the large support wall of the engine house, the working stroke of the piston would have been double this crank size.

Based upon this premise the piston would have had a working stroke of 7 feet 2 inches. Southern goes on to say that the main operating beam of this engine was very large and measured 22 feet in length. At the time of Southern's visit, the engine was working and he estimated that it was making sixteen strokes each minute, and operating at this speed the crank and flywheel would have also been revolving at sixteen revolutions per minute. He was informed by a workman that the usual speed of the engine was between twelve and twenty strokes or revolutions in each minute.

The process machinery driven by this engine gives a good indication of the great variety of items which were made at James Pickard's factory at Snow Hill. Southern says when he saw the engine it was turning a grindstone of 5 feet 6 inches in diameter with a face width of 22 inches. He estimated the weight of this at 3 tons and 4 hundredweights and this grindstone was being used for the finishing of gun barrels.

At the other extreme twenty-two individual lapping discs were being driven, which were used for grinding buckles and chaps and forty-five other spindles were also being driven and were used for brushing and polishing. One additional large lap was also being driven and was used for the dressing and finishing of circular rolls.

Southern also said the engine drove a small rolling mill where six pairs of circular rolls were used for flatting [sic] metal. It is not known what diameter these rolls were, but they had a face width of 5 inches, 7 inches and 15 inches. On his visit, Southern was also told that each workman (known as a strap or

## ROTATION BY A CRANK AND FLYWHEEL

brush) was paid 2/6d (12½p) per week

When working a twelve-hour day the engine was said to consume between 22 and 24 hundredweight of Bloomfield coal. The water needed to run the engine was lifted from a well which was situated 13 feet below the factory floor. If all the machinery listed by John Southern was operated together it adds up to ninety individual pieces of equipment, and if each machine needed one operator, the factory owned by James Pickard at Snow Hill would have been quite extensive and diverse and would have employed ninety operators. At some period between 1780 and 1801, the factory seems to have been converted into a flour mill for, in the Birmingham Directory of 1801, James Pickard's occupation is now stated as a miller maltster and coal merchant.

James Pickard died in the early years of the 19th century but his Snow Hill factory was still using the original engine until about 1879 when it was finally broken up and sold for scrap. *

The research into this manufactory and its engine was carried out in the hope of eventually having enough information to construct a model showing what the Pickard and Wasborough engine most probably looked like when it was fitted with a flywheel and crank, as patented on the 23rd August 1780. The research, so far, has revealed that the engine had a cylinder with a working diameter of 30 inches, and the piston made a stroke of approximately 7 feet. When the engine was driving the factory machinery at Snow Hill, the average rotational speed was sixteen revolutions per minute.

With Matthew Boulton and James Watt steadfastly refusing to grant James Pickard and Matthew Wasborough a licence to make their improved engines, we can have little doubt that the engine at Snow Hill worked on the same principles first developed by Thomas Newcomen almost seventy years before.

John Farey in his *'Treatise on the Steam Engine'*, which was first published in 1827, gives, a historical account of the development of the engine designed by Matthew Wasborough and James Pickard, but in this account no physical sizes are stated. However, the one piece of information shown in this book which is of inestimable value into the research of this engine is a pictorial drawing of an atmospheric engine which had been fitted with a crank and flywheel. This drawing is shown in Figure 12 on page 21.

Farey does not state this to be a drawing of the actual Pickard and Wasborough engine but, along with the other known features, the drawing does show what the Snow Hill engine most probably looked like when it was fitted with a flywheel and crank in 1780.

In the text Farey goes on to say that when an atmospheric engine was adapted to provide rotary motion the flywheel was usually 16 feet in diameter. The outside rim or annulus to the flywheels described by John Farey was made from cast iron 9 inches square.

---

* *Birmingham Weekly Mercury 23rd Oct 1886.*

## ROTATION BY A CRANK AND FLYWHEEL

This was then held into its working position by eight arms of wood.

At this point, Dr Jim Andrew of the Newcomen Society was asked for his opinion of my proposal to build a 1/16 scale model of this engine. I wanted to know from Dr Andrew whether he knew of any existing drawings or written information which could reveal what this engine originally looked like when it was converted to provide rotary motion to drive the machinery at Snow Hill in 1780. Dr Andrew has studied the Boulton and Watt papers for many years and he assured me that no information had been found about the original factory which would reveal any further engine details.

*Figure 12. A drawing taken from John Farey's book 'A Treatise on the Steam Engine' showing an atmospheric engine which has been adapted to provide rotary motion.*

## ROTATION BY A CRANK AND FLYWHEEL

He also assured me that the safest course to take was to build the model from the known details and dimensions. This would give a generic representation of what this engine most probably looked like when it was assembled in 1779 and converted to rotary motion in the autumn of 1780.

With the knowledge that Matthew Wasborough had completed the Pickard installation in 1779, a decision was taken to construct an engine based upon the standard Newcomen principle, and to construct the engine as the arrangement drawing shown in Farey's 'Treatise on the Steam Engine', Figure 12. Because the engine in Birmingham was built sixty-seven years after Thomas Newcomen had built the world's first atmospheric engine at Dudley Castle, the model was to incorporate all the latest developments which had been made to engines by the late 1770's. At this time the vacuum was still being created inside the powering cylinder by a cold water spray. By the late 1770's the mechanical parts were much better proportioned and the boilers could generate larger volumes of steam. However the thermal efficiency of the engines, even over such a long period, had never been improved: the operating cylinders were still being heated by the incoming steam from the boiler and then cooled by the condensing water spray on every stroke of the engine.

# CHAPTER 3

# CONSTRUCTIONAL DETAILS.

**The Component Parts of a Rotary Atmospheric Engine**

At this stage it is intended to describe the main component parts of the engine which was assembled at Snow Hill in the sequence in which the engine operated, when it was providing the rotary power to the process machinery at James Pickard's Birmingham Manufactory in 1780.

When each component and its function has been described the details of how each part was made by the craftsman of the eighteenth century will be given.

Following the description of how the early craftsman made this engine in 1779, I will describe how I made the more interesting parts of the model of this engine at 1/16 the original size. Figure 13 shows a front elevation of the completed model of James Pickard's engine.

*Figure 13. A front elevation of the model showing how Matthew Wasborough's and James Pickard's engine most probably looked when it had been fitted with the crank and flywheel in 1780.*

## CONSTRUCTIONAL DETAILS

### Starting the Engine from cold with an empty Boiler

The first operation would have been to remove the clamping ring which was holding the wooden cover used to seal the inspection hatch onto the boiler. With this cover removed, water could have been pumped through this opening and into the boiler until the correct working level had been reached. To bring a boiler of 13 feet in diameter to the correct level, 3,500 gallons of water would have been needed. With the level of the water correct, the inspection cover could be firmly replaced and the fire beneath the boiler shell could be kindled. Figure 14 shows how the inspection cover fitted on to the hemispherical dome of the boiler. This inspection hatch also performed two other very important functions.

*Figure 14. The wooden inspection cover fitted on to the dome of the haystack boiler.*

The first was in the manufacture of the boiler to allow one riveter who would enter the inside of the boiler while his partner positioned and set the rivets from the outside. When all the riveting was complete, this opening into the boiler would normally be sealed by a circular disc. The disc was usually made from wood held into position by a clamping ring and a number of holding down bolts. Wood was usually the chosen material because the moist atmosphere from the inside of the boiler caused the wood to swell and quickly created a steam tight joint. The second function was for routine maintenance, such as descaling as this could also be carried out by entering the boiler through this opening of approximately 21 inches in diameter.

### The Furnace or Fire

Unlike the furnace on Thomas Newcomen's Dudley Castle engine, where the coal was placed and burned at the back of the fire grate and the products of combustion drawn into a chimney stack positioned at the front just inside the fire door, the later furnaces only burned their coal on fire bars which extended

## CONSTRUCTIONAL DETAILS

to half of the base area of the boiler. These were the fire bars immediately behind the fire door opening. With the coal burning in this way, the hot gases were drawn backwards over the full diameter of the boiler's base. The hot gases were then drawn into a flue which extended all around the outside of the boiler shell. This was to transfer the maximum amount of heat energy into the water, before the depleted gases were drawn into the vertical chimney stack and finally discharged into the outside atmosphere. Positioned at the base of the chimney stack was a manual slide damper used to control the flow of air which regulated the intensity of the burning coal. This engine, it was said, consumed between 22 and 24 hundredweights, of Bloomfield coal in a twelve-hour period. The fire door and flues around the boiler with the slide damper can be seen in Figures 15 and 16.

After lighting, the fire would have burned intensely for at least five hours before enough steam had been raised to continually run the engine.

*Figure 15. The fire door beneath the haystack boiler.*

### The Boiler

The steam, which was required after condensation to create the working vacuum against which the atmospheric pressure of the earth could act was generated within a vessel which became known as a haystack boiler.

This name was used because of the hemispherical dome's similarity to the agricultural haystack. In some counties these vessels also became known as beehive boilers. By studying contemporary atmospheric engines used in the mid-eighteenth century which were powered by haystack boilers, a boiler with a diameter of approximately 13 feet would not have been too different from the boiler which was originally installed by Matthew Wasborough in 1779.

CONSTRUCTIONAL DETAILS

*Figure 16. A sectional view showing how the exhaust flue completely surrounded the base of the boiler. Also shown is a slide damper which was installed to control the intensity of the burning coal.*

## CONSTRUCTIONAL DETAILS

With no other reliable information available, this was the size of the boiler chosen for the model of this engine assembled by Wasborough in 1779 and fitted with a crank and flywheel in the autumn of 1780.

The complete haystack boiler at one sixteenth full size is shown in the photo-

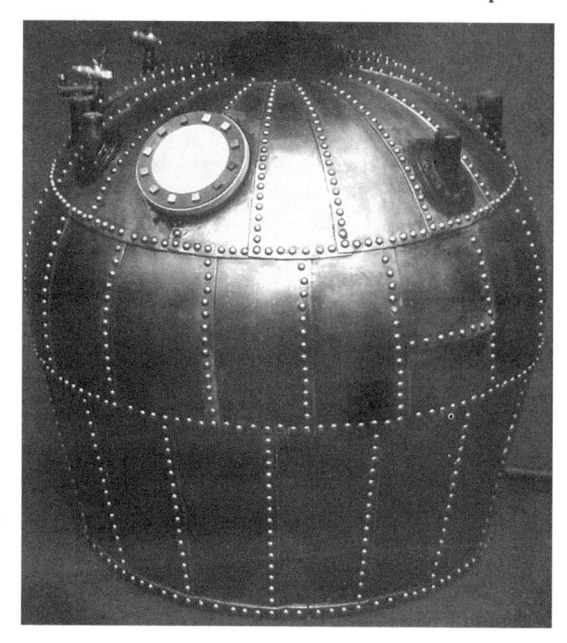

*Figure 17. The completed boiler shell, just before being permanently fixed in to its working position on the model.*

graph Figure 17, before it was surrounded within the brickwork of the model engine house.

The boiler was made from sixty-nine separate pieces of mild steel made to represent the wrought iron plates, from which the original boiler would have been made in the late 1770's. Each of these sixty-nine separate segments was individually secured together with a total of 1,749 rivets, each rivet having a diameter of 0.0625 inch. Each of these mild steel plates was beaten by hand into a wooden former, which had been made to the outside profile of a haystack boiler at one sixteenth full size. The thickness of each plate on the model is 0.020 inch, but in 1779 the boiler plates used on the Snow Hill engine would have been ⅜ inch. The only statement which seems applicable about making this boiler in miniature is that it was made with great care and patience, first developing and then accurately shaping each separate piece to the finished size, and then drilling the 3,498 holes for the rivets.

On the full sized engine at Snow Hill, the boiler plates would have been made from wrought iron. However it is not known who made the boiler of this engine

## CONSTRUCTIONAL DETAILS

from wrought iron. However it is not known who made the boiler of this engine for Matthew Wasborough in 1779. The wrought iron plates would have been made from ingots probably imported from Sweden or Russia. Charles Lloyd imported wrought iron ingots from these two countries. The iron ingots were beaten into flat sheets by his water powered tilt hammers at Burton upon Trent.

Lloyd supplied boiler plates for the construction of many of the haystack boilers which were used by atmospheric engines at this time. One notable customer of Charles Lloyd for these plates was the Boulton and Watt organisation, who used them for the construction of the Lap Engine's boiler in 1788, (the Lap Engine is described later)

The boiler, which supplied the steam to the Snow Hill engine, was most likely to have been made in the Birmingham area, because to convey such a large and heavy component over any great distance would have been very difficult more than two hundred years ago as only horse drawn carts would have been used. The making of boilers in the eighteenth century was a skilled and strenuous manual job. Each of the segments used for the dome, and also the side panels had to be made to an exact profile before drilling the holes for the rivets. It is most likely that each of the plates would have been first drawn to the full size and this drawing would then have been used as a template to cut each plate to the exact profile. These separate plates would then be beaten by a blacksmith to a shape which after riveting, would produce the well-known shape of a haystack.

When all the riveting was completed and the boiler shell was in one piece, it had to be made steam and water tight. This was achieved by first wetting the jointed faces with urine. The ammonia in the urine would promote rapid rusting of the jointed faces. On rusting, expansion would take place and help to close any gaps between the jointed faces. This was then followed by caulking. This is a process by which a blunt chisel is used all around the jointed plates both inside and outside the boiler. This closed any large gaps by burring or bending one plate on to the other. One man would have held the caulking chisel, while another man struck the chisel with a large hammer.

Finally the edges of each boiler plate would have been painted both inside and out with a blend of thin putty made from whiting and linseed oil. When the boiler was first used the heat from the burning coal and the boiling water would bake this mixture into a very hard substance which finally made the whole vessel steam and water tight. On the inside of the boiler four strong wrought iron stay bars were usually added to connect the hemispherical dome to the concave base. These additional bars were fitted to give the extra strength needed to safely withstand the internal steam pressure.

**The Dome of the Boiler with all the Fittings**

### Water level

The correct water level which had to be maintained within the boiler to raise enough steam to continually run the engine was almost 3,500 gallons. After first

## CONSTRUCTIONAL DETAILS

removing the large wooden inspection hatch, this water would have been delivered into the boiler by manually operated pumps. When the engine was in motion, the water level would be maintained by an automatic filling device, which operated through a system of levers which were all connected to a copper float resting on the surface of the water within the boiler. The water needed to

*Figure 18. The header tank used to replenish the boiler with the water which was lost by the production of steam.*

replenish the boiler after the production of steam, flowed from the header tank shown in Figure 18. This tank was positioned nine feet above the ideal water level.

The height was determined by the internal steam pressure. Nine feet would have been enough to have overcome this steam pressure and refill the boiler by the force of gravity alone. The pressure of steam required for this atmospheric engine would have been only two pounds per square inch and, the pressure created from the head of water (approximately 4½psi) would have easily overcome the internal pressure to keep the water at a constant level.

A vertical rod connected to the float can be seen passing through a steam tight seal, positioned onto the dome of the boiler.

Operated by a simple lever this rod opened or closed a cone valve to control the flow of water. Normally this valve was firmly held on to its sealing face by the lead weight shown in this photograph. When the water level dropped through the production of steam the float lowered and lifted this valve from its seat. This action allowed the water to flow from the header tank and into the boiler. Once

## CONSTRUCTIONAL DETAILS

the required water level had been reached, the float would rise and the flow of water would stop. This conserved the water within this small tank, ready for the next automatic refilling operation. In 1779, this small tank would have been cast from iron and would have held approximately thirty gallons of water. The tank on the model is made from mild steel silver soldered together and was made to appear as the original would have looked in the eighteenth century.

The flow of water into this tank was maintained from the surplus water which flowed from the top of the flooded piston. How this was accomplished will be described when details of the piston assembly are given.

**Water Level, a double Check**

A visual check on the correct water level was also made when the engine was in motion driving the factory machinery. This was performed by operating the

*Figure 19. Stand pipes with brass taps used to check the water level within the boiler*

two large brass stop taps which were fitted to the stand pipes passing through the dome of the boiler. Figure 19 shows how these taps were arranged. The stand pipes descended from these taps and were long enough to reach the correct water level. One pipe was longer than the other by about twelve inches and, the correct working water level was the midway point between these pipes. The engine man, on his routine check, would turn each of the two taps to the 'on' position.

If boiling water issued from the tap attached on to the long pipe and steam issued from the other tap, he could be confident the correct level was between the two pipes. The first engine to use a stand pipe and brass tap to gauge the water level was Thomas Newcomen's engine in 1712. This was the world's first commercial engine. In 1712 the stand pipe would have been made from lead, but by 1779 cast iron would have been used.

The taps would have been cast from brass, made to the design used for many years in the brewing industry.

Figure 21 on page 32, shows how the level was checked on the Newcomen engine, however, in 1712 only one standpipe and tap was fitted.

## CONSTRUCTIONAL DETAILS

**Boiler Safety Valves**

Two pressure release valves would also have been fitted to this haystack boiler. These two valves are shown in Figure 20.

*Figure 20. Two safety valves one valve designed to prevent the formation of a vacuum, the other valve was designed to prevent an excessive build up in steam pressure.*

The more conventional of these valves is shown positioned at the top of a tall stand pipe. This valve was designed to 'blow off' and prevent a build up of steam pressure. When the pressure became dangerously high the valve would lift from its seat and allow the steam to safely escape into the outside atmosphere. The pressure of the blow off was adjusted by sliding the lead weight along the horizontal lever shown in this photograph. This safety device would have 'blown off' when the steam pressure exceeded 2½ pounds per square inch. This was a very low pressure when compared to the boilers which powered steam engines in the nineteenth century. However, this was a large boiler with a diameter of thirteen feet and, even at this pressure, a force of twenty-one tons had to be safely contained.

The boilers used to provide the steam needed for the atmospheric engines of the eighteenth century were made to very large proportions. (The largest ever made was used at a lead mine in Derbyshire and was 20 feet in diameter).

The boilers supplying the steam to the engines of the nineteenth century were of much smaller dimensions because, the engines made after 1800 were powered by the expansive force of steam, engines powered in this way required much less steam which could be produced in smaller boilers.

These haystack steam generators of the eighteenth century were the largest vessels ever made to operate steam engines. However, the additional safety valve shown in this photograph Figure 20, was fitted to perform a completely opposite function.

When the engine was stationary at the end of a working time, and the boiler allowed to cool, the steam within this large vessel would have condensed and

## CONSTRUCTIONAL DETAILS

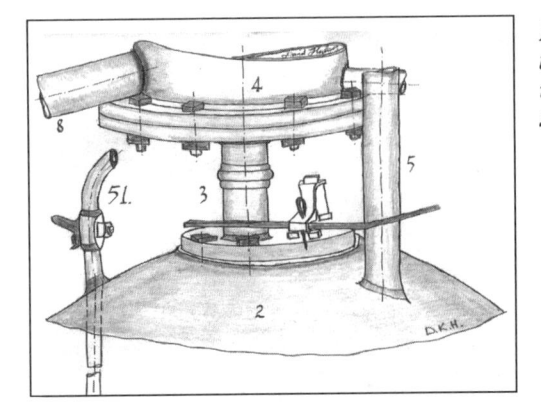

*Figure 21. The lever designed to operate the quadrant valve used to control the flow of steam from the boiler.*

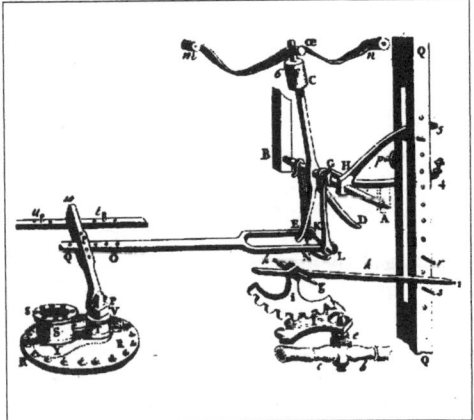

*Figure 22. A contemporary drawing of the 18$^{th}$ century showing a Newcomen type engine's valve operating levers.*

on condensing, a vacuum would have formed. If this vacuum had been allowed to develop to a maximum, the pressure exerted on the outside of the boiler shell by the pressure of the earth's atmosphere would have exceeded the safe working load for the wrought iron plates and the sides would have been drawn together. The boiler would then have been destroyed. This simple valve was held on to its sealing face by the positive steam pressure within the boiler, and when this pressure dropped by cooling and a vacuum was about to form, the seal was broken. This equalised the air pressure inside the boiler with the outside atmosphere, thus preventing any damage to the delicate wrought iron vessel. Number one valve prevented an explosion and number two valve prevented an implosion.

# CONSTRUCTIONAL DETAILS

**The Steam Transfer from the Boiler into the Powering Cylinder**
   The large quantity of steam, which was required to fill the powering cylinder of this engine, was conveyed from the boiler through a vertical standpipe which had a diameter of 6 inches. This pipe was located at the highest point on the boiler's dome. Controlling the passage of steam into the cylinder was a special valve; which was a brass quadrant and moved in a circular arc. This valve was designed to completely cover - and uncover - the full diameter of this vertical pipe. So successful was this design of steam control that its method of operation had remained unchanged for almost seventy years. Thomas Newcomen used this method to control the flow of steam into his cylinder on the world's first engine in 1712. Figure 21 shows how this valve was arranged on the Newcomen engine which, in 1712, had a boiler with a hemispherical dome made from beaten lead. Figure 22 is a contemporary drawing of the early eighteenth century, showing how the valve was operated. When the powering cylinder was full of steam, the valve, which was controlled by the levers from the up and down movement of the plug tree, was manoeuvred into the 'off' position.
   A cold water spray was then introduced and, after condensation, a vacuum was created which drew the piston to the bottom of the cylinder and a working stroke of the engine was thus completed. The longstanding reliability of controlling the flow of steam by this method was because the two mating faces were held together by the steam pressure inside the boiler. After the passage of time, the two mating faces were polished into a perfect seal. This design of valve was used throughout the eighteenth century, controlling the flow of steam into the powering cylinders of atmospheric engines and was never superseded.

**The Powering Cylinder**
   In, the Boulton and Watt collection of documents, which are held in the Birmingham Reference Library, there is a note which states the powering cylinder of the Snow Hill engine was 30 inches in diameter and that the crank had a radius of approximately 3 ft 7 inches. This radius, when doubled, would indicate that the working cylinder would have contained a piston which made a stroke of about 7 feet. These are the only two dimensions which are known about this engine. With this discovery came the realisation that a great deal of contemporary research would be needed before a model of the engine could be made.
   The earliest Newcomen engines had their powering cylinders made from cast brass up to c1730. However, by the late 1770's there had been a complete change and cast iron was now the chosen material. Therefore, there can be little doubt that the cylinder of this engine at Snow Hill assembled in 1779, would also have been made from cast iron.
   A component as large as the cylinder, would have had an overall weight of about 2¼ tons and would have been moulded and cast in a pit within the foundry floor. The internal diameter or bore of this casting would have been formed by a circular core, possibly made from mucksand. A mucksand core is made by blending together four natural ingredients: black casting sand, well rotted loam,

## CONSTRUCTIONAL DETAILS

and a generous helping of horse manure, all blended together with water into a workable paste. The core was made by coiling layers of natural hemp around a central spindle. This was then covered with a layer of mucksand and this whole procedure was repeated until the desired diameter of the core was achieved. After each layer of mucksand had been applied, the whole core was placed into an oven to dry. The final diameter of the core was achieved by hand turning on a machine, very similar to a wood turning lathe. An allowance would have been made on the diameter of the core for subsequently machining this casting to the finished size.

Why was horse manure used in the production of cores? This material is very fibrous and, when the molten iron came into contact with a core made of this material, the coarse fibres were burnt away. This allowed the molten iron to cool without creating any blow holes which would have destroyed the internal surface of the casting. The gas produced by the molten iron on cooling would escape into the core and then out to the atmosphere.

The cylinder would then have its internal diameter machined to the finished size of 30 inches on a water powered boring machine. John Wilkinson patented the principle of this machine in 1774 and a diagram is shown in Figure 23.

With the cylinder now having a finely finished internal surface, it was ready to have the piston fitted. Wilkinson always said there was no better machine available at finishing steam engine cylinders than his boring mill which was

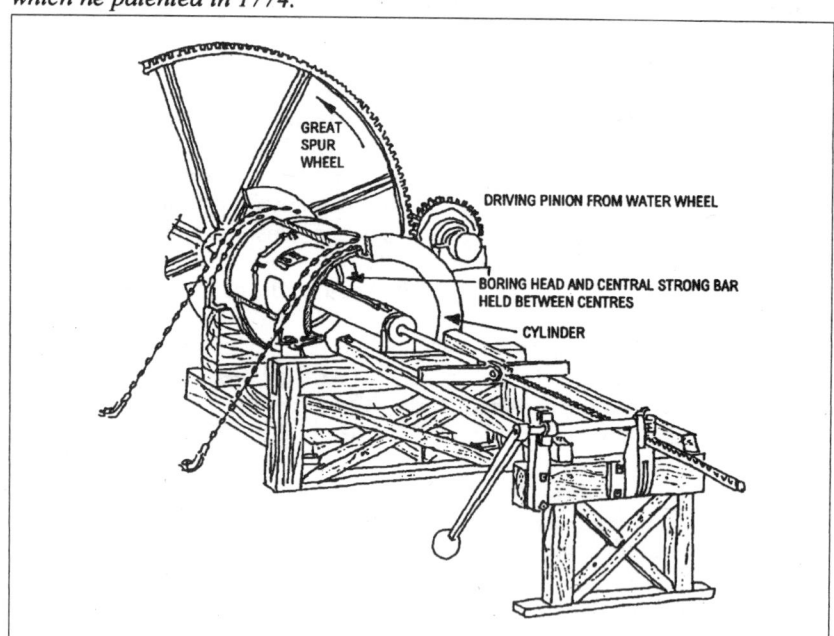

*Figure 23. John Wilkinson's water powered boring mill which he patented in 1774.*

## CONSTRUCTIONAL DETAILS

driven by an overshot waterwheel. Wilkinson told potential customers that his mill could produce steam engine cylinders with a diameter of 36 inches and a stroke of 8 feet, with a parallel tolerance of half the thickness of a well-worn shilling, (eighteenth century silver shillings are 0.020 inches thick).

**The Cylinder in Miniature**

Following my usual practice, the cylinder used on the miniature engine was again machined from a solid blank of steel. The steel blank for this component was 6½ inches in diameter with an overall length of 8½ inches and, before machining was carried out, had a weight of 75 pounds. After all the surplus metal had been turned away, more than 70 pounds of steel had been reduced to swarf. Figure 24 shows this steel held in the lathe during the machining process.

In order to make the cylinder of the model engine more aesthetically appealing I decided to make the internal diameter larger than on the original engine which was assembled at Snow Hill. The internal diameter now measures

*Figure 24. The blank of steel which was used to make the powering cylinder on the miniature engine. Turning the outside profile to size in the lathe.*

2½ inches, true to scale this should have measured almost 2 inches. At this juncture I do have to make an admission: the steel which I obtained turned out to be high tensile steel and was very difficult to machine on my 6-inch lathe. However, I had a friend with a large lathe which was used to drill a 2-inch diameter pilot hole.

With the hole drilled, the steel was returned to my lathe at home for finishing. Figure 25 shows how this was held in the lathe while the bore was finished to a diameter of 2½ inches. On completion the internal diameter was only 0.001inch out of parallel, on the whole length of 8 inches. This accuracy had been achieved on a lathe driven by a flat belt, which I have owned for forty-

*Figure 25. The steel blank having the bore for the piston turned to size.*

three years. This lathe was made c1900 by the Vernon Machine Company, Worcester, Massachusetts, U S A.

The completed cylinder assemblage can be seen on the miniature engine securely held into its working position by six long bolts with square nuts. The bolts can be seen passing through the two great sommer beams. In 1779 these beams would have been made from English Oak and were built into the strong brick wall of the engine house. Figure 26 shows this cylinder.

All the steam and water pipe fittings needed to operate this cylinder on the model were either soft or silver soldered into their working position.

The method of operation of each component relating to the powering cylinder is to be described. This description will be based upon how this engine originally worked driving the machines at James Pickard's Manufactory in the summer of 1780.

**The Piston**

The piston which was used by the latest atmospheric engines (1779) was made from cast iron and was of large proportions. If the piston used on James Pickard's engine was made to the design shown in Figure 27, it would have weighed almost 600 pounds. Shown in this photograph is a piston which was once used to power an eighteenth century atmospheric engine. This is now preserved in the South Kensington Museum. The piston shown in this photograph would have been used as cast from the foundry, the only form of machining was for the fitting of the wrought iron piston rod which would have been 2½ inches in diameter. This rod would have been secured into the piston by tapered wedges. The seal between the moving piston and the cylinder wall

## CONSTRUCTIONAL DETAILS

*Figure 26. The completed cylinder securely bolted into its working position. It is held onto the sommer beams with square headed nuts and bolts.*

*Figure 27. The piston of a Newcomen engine showing how it was attached onto the main operating chains.*

was made by coils of rope, from natural hemp. Before winding this rope tightly into the groove shown in the casting Figure 27, the rope would have been soaked with a blend of animal fats and olive oil. When the piston was assembled into the cylinder, the sealing rope would have been expanded on to the cylinder wall by tightening the eight bolts which were holding the junk ring or clamping ring into

## CONSTRUCTIONAL DETAILS

position.

So as to ensure a better seal between the piston and the cylinder wall, water was flooded over the top to a depth of approximately 8 inches. The depth of water remained constant by the fitting of an attachment onto the top of the cylinder which formed a weir. Figure 28 shows how this device was fitted and

*Figure 28. The cylinder top showing the cup which was designed to maintain a constant depth of water over the piston.*

maintained the constant depth of water over the piston. This attachment on the later engines became known as a cylinder cup. Much more water continually flowed through the tap shown in this photograph than was actually needed to maintain a seal for the piston. The surplus water then flowed over the weir and into the cylinder cup. From this cup the water drained away by gravity into the small header tank. From this tank the water was used to refill the boiler with the water, which had been lost through the production of steam. This header tank and filling device are described under the heading of 'boiler fittings'.

### Creating the Powering Vacuum

The vacuum against which the earth's atmospheric pressure could operate was created by the condensation of steam within the main powering cylinder by a cold water spray, which was applied to the underside of the piston. When the engine ceased working at the end of the day, a strict closing down procedure had to be followed. This was to manoeuvre the engine into a position from which it could easily be restarted.

The ideal position from which the engine could easily start, was with the main oscillating beam completely ' into the house'. This is the position when the piston has just completed a working stroke and is 'parked' as near as possible,

# CONSTRUCTIONAL DETAILS

*Figure 29. A valve used to control the flow of water into the cylinder opened and closed by a double sector mechanism.*

just above the boiler.

**Starting the Engine from this Stationary Position**

With the furnace burning well and the boiler producing large volumes of steam, the control valve shown in Figure 21 (page 32), would have been turned into the on position by hand. Steam would have then passed from the boiler to fill the cavity just beneath the piston.

This procedure was followed by manually revolving the flywheel of the engine and, by turning the flywheel, the piston would have returned to the top of the cylinder. This movement by the piston would have drawn with it a large amount of steam out of the boiler to fill the cylinder positioned above. With the cylinder now full of steam, the control valve shown in Figure 21 would have been turned into the 'off' position by hand. A cold water spray was then introduced into the cylinder. On contact with the hot steam the cold water caused condensation and, by condensing, a vacuum was created on the underside of the piston.

The piston was then drawn through a powering stroke to the other end of the cylinder. The cold water spray was introduced into the cylinder by again manually turning a second valve, as shown in Figure 29. This water valve was opened and closed by a double sector mechanism. This mechanism was a late improvement to the automatic operation of engines working on the Newcomen principle. The above procedure was repeated several times until the engine had gathered enough momentum to rotate freely on its own.

The two control valves could then be operated automatically, by the up and down movement of the plug tree which was connected to the main oscillating beam of the engine. This engine could now provide continuous rotary motion to an output shaft which drove all the machinery at James Pickard's factory at Snow Hill.

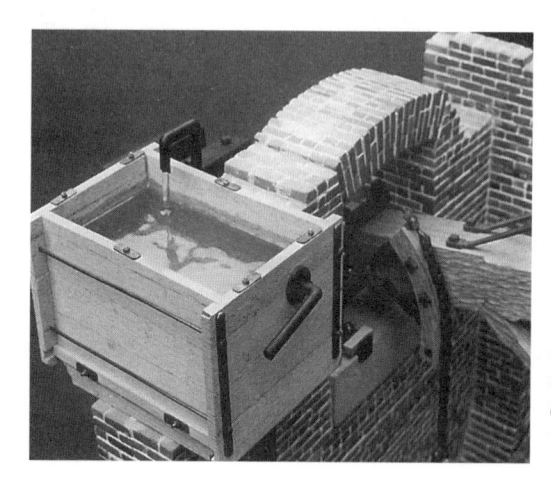

*Figure 30. A wooden header tank which was used to supply the water to the condensing spray within the powering cylinder.*

**The water Spray**

The water which was used by the condensation spray within the powering cylinder, came from the wooden header tank shown in Figure 30. This tank, shown on the model engine, is positioned much lower than would have been the case on the original engine in 1779.

Matthew Wasborough would have positioned his tank at the highest possible point, possibly in the apex of the roof. This height would have been necessary to create enough water pressure to form a good spray within the powering cylinder. The outlet pipe from this tank would need to be a minimum of four inches in diameter. To create a powering vacuum by this method would have consumed approximately six gallons of water on every stroke of the piston. This spray had to start and then stop in less than two seconds, all completed within one powering stroke of the piston.

The valve controlling this flow of water was positioned midway along the horizontal pipe from the header tank into the base of the cylinder. Once inside the cylinder the pipe turned through 90 degrees to direct the water spray vertically into the steam. This is the valve which was turned 'on' and 'off' by the double sector mechanism shown in Figure 29. Also, during one powering stroke of the atmospheric engine, the water introduced by the spray and, also the water created by the condensation of the steam, had to be drained away from the cylinder, all before the piston had completed one powering stroke. This operation had to start and then stop in less than two seconds. Another large pipe four inch in diameter was fitted into the base of the cylinder. This pipe can also be seen in

## CONSTRUCTIONAL DETAILS

*Figure 31. The accumulation of water from the powering cylinder drains into this tank and must be completed in less than two seconds, before the piston finishes its working stroke.*

Figure 26 and 33. Before discharging the water to waste, the water flowed through a one way valve, which was positioned twelve inches below the water level shown in the tank, Figure 31. Arranged in this way, the water and the air from the powering cylinder was allowed to escape.

A very important function was also performed by this one way valve - it allowed the cylinder to breathe out, but retained the vacuum which was then used to draw the piston through a powering stroke.

The overflow from the small tank shown in Figure 31 was then allowed to drain away to waste. On the earlier atmospheric engines, these non return valves became known as snifting valves, so called because the noise they made when the engine was working reminded the engine operators of an old man with a bad cold.

On studying the front elevation of James Pickard's engine in model form, the water pipes do seem odd. The delivery pipe supplying the water into the header tank is two and a half inch in diameter, whereas the water outlet pipe is four inches in diameter. The answer to this conundrum is that the water into this tank was delivered under high pressure from a force pump operated by the up and down movement of the main oscillating beam of the engine.

### The Tumbling Bob

When the atmospheric engines of the eighteenth century were working to remove flood water from the coal and mineral mines, their working speed was very low - usually the engines only completed ten working strokes each minute. However, an engine which moved at this very slow speed had no direct means of opening and then closing the two control valves which were needed to control

## CONSTRUCTIONAL DETAILS

the automatic cycle of the engine. The operation of the valve which controlled the flow of steam into the powering cylinder and also the operation of the valve, which controlled the flow of cold water into the cylinder to form the water spray,

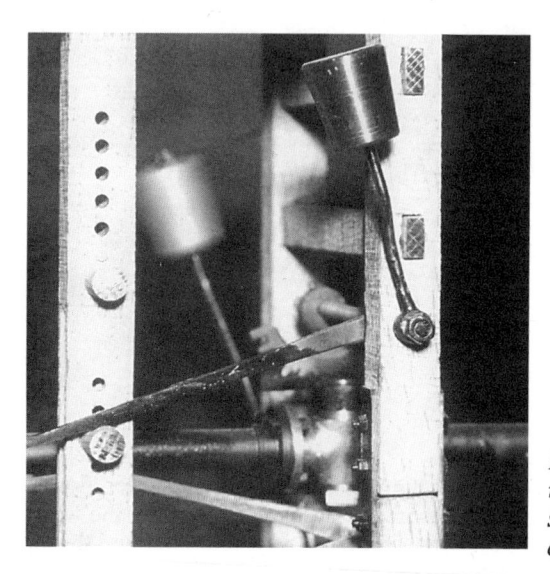

*Figure 32. The tumbling bob used to control the water spray into the powering cylinder.*

had to be turned on and off very quickly. These are two movements which could not have been achieved directly from an engine moving so slowly. To convert the very slow up and down movement of the plug tree (valve operating rod) into the snap action needed for the operation of these two valves, a clever and simple device was fitted and, when used on these atmospheric engines, became known as a tumbling bob.

The lever shown in Figure 32 was moved vertically up and down by the wooden tappets secured on to the side of the plug tree.

By the movement of the plug tree, this horizontal lever rotated a shaft on to which was mounted a vertical lever with a heavy cast iron weight secured at the highest point. When this second lever had moved slowly in an arc slightly over the vertical position, the weight tumbled by the force of gravity until the operating lever came into contact with one of the wooden tappets. This simple mechanism had converted the slow movement created by the main beam of the engine into the very quick action so important for the automatic operation of an atmospheric engine.

Both the steam admission and the water spray valves were controlled by tumbling bobs. The volume of steam needed to fill the powering cylinder was controlled by adjusting the leather strap shown in Figure 33. So successful was the tumbling bob at controlling an atmospheric engine that it remained

## CONSTRUCTIONAL DETAILS

unchanged for almost seventy years. Thomas Newcomen was the first engineer to use this method on the Dudley Castle engine in 1712.

*Figure 33. The second tumbling bob used to control the flow of water into the powering cylinder.*

**CHAPTER 4**

# THE MAIN STRUCTURAL PARTS OF THE ENGINE

**The Main Beam**

Supported by two large plummer block type bearings and securely held within the strong gable end wall of the engine house, is the main operating mechanism of the whole installation. This mechanism became synonymous with this type of prime mover worldwide, and they are all now classified as beam engines.

John Southern, on his visit to see the Snow Hill engine, stated that this beam was twenty-two feet long and was made of generous proportions.

With no other dimensional information available, the beam on the model engine was made to represent this length, with a twenty-four-inch square section. This cross sectional size was taken from an engine designed by James Watt which was also assembled in 1779 only two miles from the site of the Pickard engine at Snow Hill. The Watt engine was of course the Smethwick engine which had a powering cylinder of thirty-two inches in diameter, almost identical to the size used on the Pickard engine which was powered by a cylinder thirty inches in diameter.

The main beam in 1779 would have been made from English oak; on the model engine it is made from Japanese oak. When making models where a large amount of timber is used, a substitute to the original wood has to be found. This is because whilst it is easy to reduce many of the original components size to the chosen scale of the model, what must also be achieved is to reduce the size of the growth rings as far as is practicable, and Japanese oak is the ideal substitute for English oak. In every other feature its appearance is the same, but with a much finer texture and closer growth rings. In the eighteenth century this large baulk of English oak would have been cut to the finished size by men wielding adzes. This appearance has to be reproduced on the model by a hand-held chisel with a radius ground on the cutting edge.

Figure 34 shows this finished beam held into its working position, projecting through the gable wall of the engine house.

Secured at right angles onto the main oscillating beam is another strong piece of timber. Again, this would have been English oak and was positioned directly over the centre line of the powering cylinder. At the outer edge of this attachment was formed a radius. This radius is equal to the distance from the pivot point of the main beam to the centre line of the powering cylinder.

The generally accepted name for this attachment is an arch head, but sometimes it is also referred to as the horse's head. These were used to compel the operating chains to move up and down in a straight line.

A study of Figure 34 will reveal how this mechanism worked on the single acting engines of the eighteenth century. Weights which were attached onto the other end of the beam drew the piston back to the top of the cylinder to enable another powering stroke to begin. How these additional weights were specified

## THE MAIN STRUCTURAL PARTS OF THE ENGINE

*Figure 34. The main oscillating beam of the engine showing the strong tension chains around the arch head.*

will be described later.

Also, shown in Figure 34 is a smaller arch head, which was designed to provide a parallel action to the up and down movement of the plug tree. Plug trees were designed to operate the two control valves which were positioned at the base of the cylinder - these valves were used to control the engine which could then run unattended.

**The Arch Head Chains**

The connection between the moving piston and the arch head was made by the two strong chains which are shown in Figure 35 on page 46. The design for these chains was again taken from the ones which were specified for the Smethwick engine and the photograph shows how the chains would have looked when the Snow Hill engine was completed by Matthew Wasborough in 1779. Each length of chain on the model is made from one hundred and six separate parts, with each link being individually forged by hand. In 1779 the chains used on the Smethwick engine would have been made from wrought iron by Samuel Rudge and Co., Samuel Rudge charged the Birmingham Canal Navigation Company, £25.9s.5d for these two chains. Because the Snow Hill engine was of similar proportions to the Watt engine, and the two engines were assembled only two miles apart, I think it is reasonable to assume that Samuel Rudge could have also supplied the Snow Hill engine with its arch head chains, particularly as this design of chain appears on many of the single acting engines

## THE MAIN STRUCTURAL PARTS OF THE ENGINE

*Figure 35. A close view of the tension chains and the tap used to control the flow of water over the piston.*

of the period. The additional chain which is shown in this photograph was used to lift the plug tree, while a weight secured to the bottom of the plug tree then pulled it down again. Here I have used a simple chain with a riveted construction. This was chosen because the only force on the chain was the weight of the plug tree which needed to operate the two control levers.

### Rotary Motion from a Single Acting Engine

When the atmospheric engines of the eighteenth century were used to draw the flood water from the coal or mineral mines, the heavy pump rods which were attached onto the main oscillating beam were heavy enough to draw the piston back to the top of the cylinder to enable another powered stroke to commence. However, when a single acting engine was used to provide rotary motion there were no heavy rods in use to complete this simple task. The engineers had to devise another method of returning the piston to the top of the cylinder before another powered stroke could begin.

Their solution to the problem was to make a simple calculation for the force which would be exerted by the powering cylinder. An assumption was made that a vacuum of thirteen inches of mercury could be achieved on the underside of the piston, which would give an effective force of 6½ pounds per square inch on the top of the piston. In the case of the Pickard engine, which was using a cylinder of thirty inches in diameter, this force would have been 4,595 pounds. This is the force which the piston would have been able to exert as a direct pull onto the arch head chains. This calculated figure was then divided by two, which then equalled 2,297 pounds. This was the weight the engineers now considered had to be securely attached onto the other end of the main beam to actually haul the piston back to the top of the cylinder before another powering stroke could begin.

## THE MAIN STRUCTURAL PARTS OF THE ENGINE

*Figure 36. Cast iron attachments with a combined weight of 1.5 tons used to return the piston to the top of the cylinder to allow another powered stroke to start.*

Figure 36 shows the two weights which were secured onto the main beam to perform this important function.

I have shown two weights because one heavy weight would have been attached as close as possible to the connecting rod pivot the ' little end bearing,' and a lighter weight could have been easily moved along the beam before being finally cramped into the ideal working position when the engine was running smoothly. On some installations, the engineers stated that if a heavy cast iron connecting rod was used, these weights might not be necessary. What is not known is whether any installation ever used a cast iron rod to keep a single acting engine reliably rotating.

The weights which were used were held in their working position by strong wrought iron clamps and long bolts. In 1779 these weights would have been made from cast iron. Figure 36 also shows the little arch head which was used to provide linear movement for the water feed pump. This was a bucket pump and is shown in Figure 37. Bucket pumps worked by lifting a column of water which was constrained above the piston within the pump barrel. As the piston of this pump was lowered by the action of the main oscillating beam, the water flowed freely through two non return valves.

On the rising movement of the beam, these valves were closed by the weight of the water trapped above the piston. When the engine had completed one powering stroke, the water above the piston had been forced to rise through a delivery pipe and into the wooden tank which was positioned high in the roof of

## THE MAIN STRUCTURAL PARTS OF THE ENGINE

*Figure 37. The lift pump which was used to deliver water into the wooden header tank. From this tank the water flowed on to the top of the piston and, also created the water spray inside the powering cylinder.*

the engine house. From this tank the water flowed by gravity and was used to replenish the boiler. The height of this tank was enough to create the water pressure to form the condensing spray within the powering cylinder. On his visit to see the engine, John Southern stated that the condensing water was drawn from a water course thirteen feet below the factory floor.

**The Main Pedestal Bearing Blocks**

When these atmospheric engines were being used in the eighteenth century, the main bearings used as the pivot for the main oscillating beam would have been of a special design. Because of the great weight of the beam and all its fittings, the engineers never considered it necessary to use bearings with bolted on caps. The great weight was always enough to prevent the beam from lifting out of its working position. However this is another instance where a model has to be different from the prototype. Traditional bearings have been used with split phosphor bronze shells. This is because the weight of the beam in miniature was not enough to hold the beam into the working position by the force of gravity alone.

**The Crank and the Flywheel**

Figure 38 shows, for the first time in recorded history, a crank and flywheel providing rotary motion directly from a single acting engine.

## THE MAIN STRUCTURAL PARTS OF THE ENGINE

*Figure 38. The gable wall showing the flywheel and the crank. Also shown is the wooden tank high on the wall.*

For the past seventy years, engineers had strived for a solution to the problem of how to harness the erratic movement from the mine pumping engines to directly produce rotary motion. All the methods tried before 1780 had taken into account the engine's unpredictable nature. Now, for the first time, the crank had solved all the engineers' problems.

The unpredictable movements of the atmospheric engine were now at an end and the crank was compelling the engine to make working strokes of an equal length. When Matthew Wasborough's flywheel had been fitted, the fluctuations during each revolution, were greatly reduced.

### The World's first Connecting Rod

The design of the main oscillating beam of the Snow Hill engine differed from the usual practice used in the construction of atmospheric engines. Only one arch head was fitted. The other end of the beam had been modified to take a new design of attachment - a pivoted bearing was bolted into position on the end of the beam, on to which was attached a long rod, now known universally

## THE MAIN STRUCTURAL PARTS OF THE ENGINE

as a connecting rod. The design of this pivot, which was fitted onto the model, was taken from John Farey's book *'A Treatise on the Steam Engine'*.

What is difficult to imagine in the year 2001, is that this bearing holding the connecting rod into its working position is the world's first small end bearing and this name has been associated with connecting rods ever since. The upper end of the connecting rod was first attached on to the small end bearing before being connected to the crank twenty-three feet below. John Southern on his visit to see the engine at Snow Hill, said that this crank had a radius of three feet and seven inches. The joining of the connecting rod onto the crank was also a new innovation because, what we are looking at here is a rotary joint with split bearings and known to this day as a big end bearing.

The connecting rod on the Pickard engine was very large. The distance from the centre of the little end, to the centre of the big end, would have been twenty-three feet. This rod would have been made from two lengths of English Oak, securely held together with wrought iron straps and bolts. Two pieces of timber would have been used because, in 1780, one straight piece of wood more than twenty feet long would have been very difficult to obtain. Figure 39 shows how the big end bearing and the connecting rod were secured onto the crank.

What is not known is what the crank on the Pickard engine looked like, or what design he used for the construction of the big end. The design used on the model engine was taken from contemporary engine drawings of the late eighteenth century.

### The Flywheel used with a Crank

No actual dimensions are stated in Matthew Wasborough's patent of 1779 for the addition which he fitted onto the output shaft of his engine at Bristol, referred to in his patent specification as a *'Fly'*. However in John Farey's book it is stated that flywheels when used on single action engines were usually sixteen feet in diameter. Farey goes on to say that the outside rims of these flywheels were made from cast iron and were heavy, and that these heavy rims were held into their working positions by strong arms of wood.

### Construction of the Flywheel

From John Farey's written evidence, the most likely method of construction for this flywheel would have followed the tried and tested technology of the late eighteenth century.

The methods used for the construction of waterwheels would have been used. The constructional details for the model were obtained by travelling around Staffordshire and Derbyshire measuring and photographing as many water wheels as possible. Fortunately in these two counties there is an abundance of restored water driven mills.

The design of a wheel which was finally settled upon was taken from the Cheddleton Flint Mill in Staffordshire. At this mill, two large wheels can be seen driven by the River Churnet. These wheels provide the rotary drive to the grinding pans, which were used to prepare the flint before being used in the

# THE MAIN STRUCTURAL PARTS OF THE ENGINE

*Figure 39. The connecting rod with the big end bearing fitted onto the crank. Also shown is the central hub and the balance weight.*

pottery industry. The central hub of one of these wheels can be seen in Figure 40. Also clearly shown in this photograph is the method which was used to secure the central hub onto the driven shaft. On the water wheels at the Cheddleton flint mill there are ten arms of wood holding the paddles into their working position. Because the flywheel of Matthew Wasborough's engine was four feet less in diameter than the water wheel at Cheddleton, I decided that eight arms of wood would look more in proportion.

The central hub of the finished flywheel fitted onto the model engine can be seen in its working position in Figure 39. The outside rim of the flywheel on the model engine is larger in diameter than the sixteen feet specified by Farey. On such a large engine as James Pickards eighteen feet in diameter seemed more appropriate, but I did follow his recommendations and made the outside rim heavy. The rim at the scale of the model, measures ¾ inch square section. This would have equated to the full size engine having a flywheel with a weight of 6½ tons.

Following my usual practice, the flywheel on the model, was again machined from a solid piece of mild steel. Before any machining had taken place, this blank was 14 inches in diameter and 1 inch in thickness. Too big for my lathe at home, I used a friend's large centre lathe to machine this wheel to a diameter of 13½ inches.

51

# THE MAIN STRUCTURAL PARTS OF THE ENGINE

*Figure 40. The central hub of a waterwheel, a photograph taken at the Cheddleton flint mill, in North Staffordshire.*

To greatly improve the smoothness of the rotary motion from a single acting atmospheric engine, John Farey also stated that the combined weight of the connecting rod and the crank must also be balanced. His recommendation was that an adjustable weight must be securely bolted onto the opposite side of the flywheel to counteract any out of balance forces. Figure 39 shows how this has been achieved on the model engine. Two weights are shown clamped onto one of the wooden arms of the flywheel.

The final radial position of these weights was settled upon when the perfect balance was achieved. Once this position was found, the two pins were firmly tightened to clamp these weights into their final working position.

**Protected by the Law.**

Hero's whirling aeolipyle was the first recorded experiment which achieved rotary motion by the power of steam in 100AD. However, these early philosophers only harnessed the power of steam to satisfy their intellectual needs in a small experimental way. Almost seventeen hundred years had to pass before we could see a factory being driven for the first time directly from the properties of steam. James Pickard and Matthew Wasborough must have been justifiably proud to have been the first two men to have achieved this much sought after goal.

In Figure 41, Pickard can be seen in stately dress with Wasborough reading through the final draft of the patent number 1263, which was taken out on the 23rd August 1780 for their sole rights to this invention. This patent became a lawful document on 9th December 1780. The Patent reads: '

For a new method of applying steam engines, commonly called fire-engines to the turning of wheels, whereby a rotary motion or motion round an axis, is performed and the power of the engine is more immediately and fully applied (where motion round an axis is required) than by the

# THE MAIN STRUCTURAL PARTS OF THE ENGINE

intervention of a water wheel'.

James Pickard and Matthew Wasborough must be given the credit for being the first men to provide rotation directly from a steam engine by the use of a crank and flywheel.

The patent finally expired in 1792, but during this twelve-year period James Watt used the invention of the sun and planet to achieve rotary motion on his engines. However, soon after 1792, all the rotary engines made by the Boulton and Watt organisation were supplied to their customers fitted with the simple cranks. The engine by James Pickard and Matthew Wasborough can claim to be the first engine in the world to achieve rotary motion by the use of a crank and flywheel.

**The research and the engine in miniature are now complete.**

The details needed to make a model of the Pickard engine were gathered together over a period of almost twenty years. While I was researching and making the other engines describes in this series, I was forming a picture of what this engine most probably looked like when it was built in 1779. No detailed drawings or illustrations of this engine are known to have survived.

The finished model is shown in Figure 13 and is 36 inches long and 24 inches wide and has a height of 34 inches. There are 9,200 reduction fired red clay bricks in the wall, which is shown holding the main oscillating beam into its working position. In order to cover the engine house floor, 1,400 engineering blue bricks have been used. The haystack boiler is 8 inches in diameter and 8 inches tall. This boiler is made from sixty-nine separate segments of mild steel held together with 1,749 rivets. No part of the model is made from commercially available items, the flywheel is 13½ inches in diameter and there are 367 hand made square headed pins and nuts with washers. The wooden water tank is made from beech and structural parts of the engine are made from Japanese oak. The actual construction time for the Pickard and Wasborough engine in miniature was 2,800 hours.

THE PATENT IS STUDIED IN 1780

*Figure 41. James Pickard and Matthew Wasborough reading through the patent which was granted for the crank on the 23$^{rd}$ August 1780. These two men were modelled in bone china to the same scale as the engine in miniature by Steve Leadly a sculptor from Royal Doulton.*

# AN INTRODUCTION TO THE DOUBLE ACTING ROTARY BEAM ENGINE

**James Watt's Engines in Cornwall**

The first atmospheric pumping engine made to the design of Thomas Newcomen to draw water from a mineral mine in Cornwall was set to work at Wheal Fortune, a copper mine on the boundary between St. Hilary and Ludgvan parishes. This engine has been authenticated as working in 1720 and was followed by an engine constructed in 1725 at Wheal Rose, Chacewater, near Truro. In the following two years, a further two engines were built - one at Wheal Busy at Chacewater and the other at Polgooth, near St. Austell. All of these engines were erected by Joseph Hornblower, a Staffordshire man who had made Thomas Newcomen's acquaintance during the pioneering days of Newcomen's engines. For the next fifty-seven years, the Newcomen type of engine remained unchallenged for draining water from the mines, even though it had a prodigious appetite for consuming coal. This had to be accepted if the vast mineral wealth below the natural drainage, or adit, was to be mined safely. As the mines became deeper in the never ending search for minerals, the Newcomen engine was found to lack the power to draw the flood water from the greater depths.

It was thought that about seventy-five Newcomen engines were built in Cornwall by 1777, and it was in this year that James Watt and Matthew Boulton erected their first engine in that mineral rich county. Watt's engine was assembled at Wheal Busy, Chacewater, and was a single-acting engine with a powering cylinder of 30 inches in diameter.

Although the engine was still powered by the atmospheric pressure, this time the vacuum applied to the underside of the piston was created by the condensation of steam inside James Watt's patented separate water cooled condenser. So successful was the Watt engine that by about 1783 only one Newcomen engine was said to be left in the whole of Cornwall, and even this lone survivor was not at work.

The rapid demise of the Newcomen engine in Cornwall was due to the fact that the pumps driven by these engines had reached the greatest depths from which they could successfully draw the flood water from the mines, together with the nearest source of the coal for the furnaces being in South Wales.

The Watt engine was the answer which all the mine captains had been waiting for - it was far more powerful, and it consumed much less coal. In other parts of the country, the early Newcomen engines were drawing flood water from coal mines where the cost of coal was not such an important factor and the coal mine was not usually as deep as the tin and copper mines in Cornwall.

Boulton and Watt only sold their engines under the terms of an annual payment to them of one-third of the cost of the coal savings over that of a standard Newcomen engine of equivalent power.

## THE DOUBLE ACTING ROTARY ENGINE

The reward to the patent holders and the economic value of the separate condenser to Cornwall was typified by the case of the Great Consols Mines in Gwernap, where seven Newcomen engines in 1778-79 consumed 19,086 tons of coal, whilst the five Boulton and Watt engines which had replaced them by 1783 consumed less than one third of this ( 6,090 tons ) during a similar period. This saved £9,097 per annum for the mine owners and Boulton and Watt agreed to an annual payment of £2,500 for these five engines.

The last engine to be built based upon the Newcomen operating principle was an engine built by Joseph Thompson, the oldest son of Francis Thompson. As late as 1825, he installed this engine at the Magpie lead mine at Sheldon near to Bakewell in Derbyshire. This engine had a 42 inch diameter cylinder with a stroke of 9 feet. This engine at the Magpie mine only worked until 1840, when it was superseded by a more powerful Cornish Engine.

However an engine which was built in 1797 to remove the flood water from the Elsecar Colliery near Barnsley, is now the only atmospheric engine to survive which is preserved on its original site.

# CHAPTER 5

# THE BOULTON AND WATT LAP ENGINE OF 1788

**Double Acting Cylinders**
So successful was the separate condenser in reducing the amount of steam needed to operate these engines that the age of cylinders powered in both directions, or double acting cylinders, was now dawning. These engines were needed to provide the rotary power for industries which had previously been reliant on flowing water or horse power. With the introduction of double acting cylinders, the next engine in this series of early steam engines can now be considered. Possibly the most famous of all Boulton and Watt engines ever made was the Lap Engine of 1788.

With the introduction of the rotative beam engine, the Cornish mining industry was to have a further invasion of Boulton and Watt engines, as the mined ores could be hauled safely to the surface by the new engines which replaced the horse whims. H W Dickinson and R Jenkins record in their book, '*James Watt and the Steam Engine*', that sixty-one Boulton and Watt engines were erected in Cornwall between the years 1777 and 1800, with eight of the rotative engines being powered by double acting cylinders.

**The Lap Engine**
James Watt designed and built this beam engine in 1788 to produce rotary motion. The engine was powered both by a vacuum and by the expansive force of steam, with the vacuum being created by the condensation of steam inside Watt's patented water cooled separate condenser. The vacuum was first directed to the top, and then the underside of the piston. This combination of both a vacuum and low pressure steam produced a continuous output of energy which powered the engine.

By 1788 many other beam engines had been adapted to provide rotary motion, but they had not been designed and built for this exclusive purpose and they were mainly adaptations of single acting pumping engines.

This rotative beam engine was erected in July at Boulton and Watt's Soho Manufactory in Birmingham, where it became known as the Lap Engine because it was used to lap and polish small manufactured components such as the large buckles used on court shoes. The operators polishing these small items were generally referred to as 'toy makers'.

The Lap Engine continued to drive the factory machinery until 1858 when the Soho Manufactory ceased working. The engine is now on display at the South Kensington Science Museum London and is shown in Figure 42, on page 60. The framework of the engine was built entirely from wood, all of which was held together by wrought-iron straps and bolts. The engine had a double acting cylinder of almost 19 inches in diameter with a working stroke of approximately 4 feet. The engine was the first in the world to have its rotational speed controlled by a centrifugal, or Watt, governor. The large flywheel with three

# THE BOULTON AND WATT LAP ENGINE OF 1788

*Figure 42. The Lap Engine in the Science Museum, South Kensington London.*

hundred and four wooden teeth drove more than forty lapping and polishing machines at the Soho Manufactory and the rotary power was transmitted to each individual machine by belts and pulleys.

The Lap Engine was one of the first rotary beam engines to have its power output rated in horsepower. This power was calculated by James Watt to be 10 normal horse power, with 14 indicated horse power produced when the engine was in motion.

When the Soho Manufactory ceased working the engine was put into storage where it was to remain until 1861. Matthew Piers Robinson Boulton then presented the engine to the Patent Museum in London, where the engine remained on display until 1926. Unfortunately, the engine was not reassembled in the Patent Museum in the way that it had originally been used at the Boulton and Watt Manufactory. The engine, with its large flywheel of almost 16 feet in diameter, was assembled like many later nineteenth-century engines with the lower half of the flywheel below the factory floor level. The engine erectors were unable to dig a pit because the engine was assembled on the first floor of the museum and the parts below the floor level were therefore put into storage.

The engine was transferred in 1926 from the Patent Museum to the South Kensington Museum, but through the years some of the parts which had been placed in storage had become mislaid and the lower half of the flywheel is now

## THE BOULTON AND WATT LAP ENGINE OF 1788

actually made from wood. At least the engine is complete again below the working floor level, even though the engine is still not assembled how it had been used at the Boulton and Watt Manufactory. Only the engine is displayed in the South Kensington Science Museum and there are no remains of the original waggon boiler which supplied the steam to power the engine. This fact was discovered from two drawings dated 29th July 1788, and fortunately these drawings show the Lap Engine residing in the same building side by side with the boiler - an unusual feature because at this date most engines and boilers were sited in separate buildings.

A great effort has been made to create the model of the Lap Engine as it would have appeared in 1788. The miniature engine has been constructed from the same materials as would have been used by the engineers in the eighteenth century. This project took five years to research and construct in miniature.

### An Introduction to the Miniature Lap Engine

Many people derive great pleasure from the making of miniature mechanical devices and perhaps I could be counted as one such person. My first attempt as a miniaturist was in the making of a four-column beam engine which was powered by means of a vertical gas boiler. The drawings were purchased from A J Reeves of Birmingham and it was decided not to use the commercially available castings but instead to fabricate the parts from solid metal in my home workshop. Through some good fortune, this small model was successful and so paved the way for further and more ambitious plans. I had been searching for many months for a challenging subject to model and in June of 1978, I visited a Design Engineering Exhibition held at Olympia, London. As there was some time to pass before returning to Stoke-on-Trent, I revisited the South Kensington Science Museum in order to take another look at James Watt's Lap Engine which had always held a fascination for me. On studying the engine in more detail, I soon decided that my next project was to be the modelling of this early Boulton and Watt engine.

The Lap Engine is displayed without its boiler or engine house and enquiries were made to see whether there were any detail drawings or artists' impressions of the engine which were available to the general public. The only drawings which were available to me at this stage were two arrangements produced in 1950 by the staff of the museum. However, these drawings did not have any dimensions which would be needed to make an accurate model. Neither was there any information on these drawings as to how the engine had been originally installed at the Soho Manufactory in Birmingham where the engine worked from 1788 until 1858. This made me realise that a great deal of research would be needed before a start could be made on my chosen project.

Two weeks later I returned to the Science Museum with my camera and a wooden rule and on this visit more than one hundred black and white photographs were taken of the engine. The wooden rule was placed on every component of the engine as each photograph was taken so that I could determine the exact size of each part. Gleaning information in this way greatly reduced the

## THE BOULTON AND WATT LAP ENGINE OF 1788

amount of detail drawings which would be needed for the actual construction of the miniature engine. However, I still required many more visits to the Science Museum to check on dimensions and, of course, to obtain more details.

By November 1980, the model had been completed and I was very pleased to exhibit it at the Midlands Model Engineering Exhibition organised by TEE Publishing at the Bingley Hall in Birmingham where I received first prize and the Modelcraft Cup. In January 1980 I exhibited the model at the Model Engineering Exhibition in London and again I was very fortunate to win first prize and the Championship Cup. The model at this stage is shown in Figure 43.

*Figure 43. The original model of the Lap Engine which was inspired by the display in the Science Museum, South Kensington, London. Unfortunately, the display in the museum does not include a boiler and my beehive boiler was not correct.*

However, my triumph was short-lived, because in the March edition of *Model Engineer* it was pointed out to me that I had made several constructional errors. After this report, Professor D H Chaddock wrote to me and suggested that I should contact Mr Michael Wright at the Science Museum who had found two original drawings dated 29th July 1788. These drawings clearly show that a waggon boiler was originally used and not the haystack boiler which I had made and fitted to my model. With these drawings, my errors could now be corrected and I set to work and completely dismantled the model to rebuild all the component parts of the first model of the Lap Engine into a sectional view of the original engine house. The details of the original engine house were also shown

# THE BOULTON AND WATT LAP ENGINE OF 1788

*Figure 44. The re-built model of the Lap Engine after it had been built into a sectional arrangement of the original engine house at the Soho Manufactory. This time the boiler was correct.*

on these two drawings. Both of these drawings were dated 29$^{th}$ July 1788 and signed by Matthew Boulton.

Nine months later, the model re-emerged as an exact miniature of the Boulton and Watt Lap Engine that had been built originally at the Soho Manufactory, Birmingham in 1788. Hoping that everything should now be correct, I submitted several photographs of the finished model for verification to the staff of the Science Museum. In the reply a comment was made that the flywheel spokes were too thick at the miniature's scale and that 1/16inch must be removed. And also that nuts and bolts should not have been used on the counter-shaft couplings as wedges would have been use in 1788, also the outer edges of the main shaft bearing brasses should be octagonal and not hexagonal as I had made them.

Finally, the hacksaw shown on the workbench was of the wrong type to be representative of 1788 and should be a Lancashire hacksaw, Figure 45.

The above points were corrected immediately and at last I was satisfied that the original factory site setting of the Boulton and Watt Lap Engine had been faithfully recreated. In November 1981 this miniature was exhibited at the Midlands Model Engineering Exhibition where it gained first prize and the

61

# THE BOULTON AND WATT LAP ENGINE OF 1788

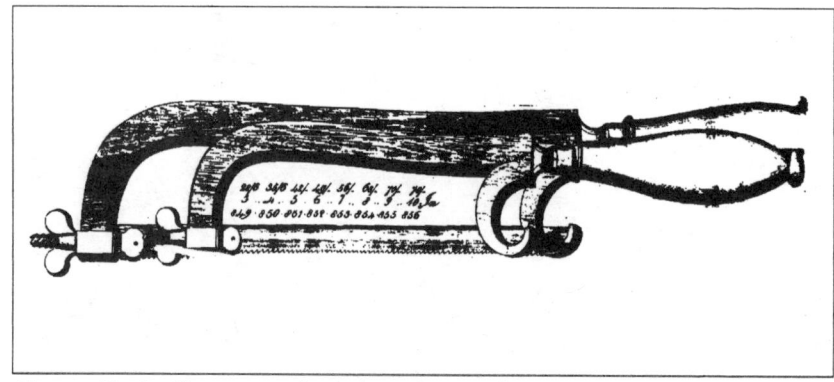

*Figure 45. An illustration from a catalogue of Lancashire bow saws which were made in the eighteenth century by R. Timmins & Sons*

Modelcraft Cup.

Having introduced the model of Boulton and Watt's Lap Engine, I will describe the evolution of the engine and how it was made by James Watt in 1788 and also, how many of the parts were made in miniature.

The same principles of construction were followed as for the small beam engine and, where it was practical, all materials used on the model were the same as those used in 1788 for the full size engine. More than 11,000 ceramic bricks were produced for the engine-house. The bricks were made by hand and were fired in a homemade kiln which gave the tiny bricks the appearance of having been made two hundred years ago. The foundation of the building was made from 1,500 hand-cut sandstone blocks. On counting all the bricks, nuts, pins, washers and hand forged panel pins, more than 36,000 separate parts have been made individually.

The toothed flywheel was the most time-consuming single item to be made and was constructed from 509 separate parts.

**The Boiler of the Lap Engine**

In his quest for higher efficiency, James Watt used a new type of boiler to power his improved beam engines. The type of boiler used to power the Lap Engine was called a 'waggon boiler' because of its rounded top, which looked similar to the covered waggons used by travellers and gypsies from the sixteenth century. Before the discovery of James Watt's separate condenser, atmospheric engines were mainly powered by haystack boilers, but these boilers were inefficient when they were compared to the more modern waggon boilers, and consumed more coal for the same quantity of steam produced. In some parts of the country, haystack boilers were also known as beehive boilers and a boiler of this type is still to be seen at Cheddleton Flint Mill, Staffordshire.

The overall size of both waggon and haystack boilers was determined by the power output of the engine - each cubic foot of operating cylinder volume had

## THE BOULTON AND WATT LAP ENGINE OF 1788

a known surface area which had to be exposed to the heating element, namely burning coal. The Lap Engine's boiler was of a very small size and volume compared to most subsequent boilers. This reduction in size was the result of using a central flue construction. The central flue greatly increased the area exposed to the burning coal and the hot exhaust gases, before they were drawn into the chimney stack and into the outside atmosphere.

The area exposed to heat on the engine's small boiler was eighty-five square feet, this area was more than adequate if Watt's simple rule of 'eight square feet of heating surface for every cubic foot of operating cylinder volume' was used. Watt formulated this rule in 1783, for single action engines and was multiplied by two for the Lap Engine, as this engine was powered by a double acting cylinder.

A much simpler construction was used for subsequent boilers, the extra heating area being gained by increasing the boiler's overall size in order to achieve the required surface area without any need for the complicated central flue. Difficulties were created by the central flue boilers as the watertight joints were awkward to achieve in such a restricted space. The internal restriction of the flues made cleaning difficult on these wrought-iron boilers. Many boilers powering engines after 1788 were of larger dimensions and were constructed without the central flues.

Some of the contemporary references show waggon boiler installations with their tops covered to conserve the heat. In most cases, this covering consisted of a 3 inch layer of horse or cow dung and a 1 inch thick layer of lime mortar in turn this was covered by two layers of bricks. However the Lap Engine's installation drawing clearly shows the boiler top uncovered and there is no indication of any lagging. The two drawings, Figures 46 and 47, have been reproduced from the original sketches signed by Matthew Boulton and dated 29th July 1788.

One possible explanation as to why no lagging was used on the Lap Engine's boiler is that it was an unusual installation with the engine and the boiler in the same building. Much of the recorded information for this date shows the boilers and the engines separated by a vertical wall, whereas when a boiler was placed outside the building, any heat loss would be a problem - so with the advent of the small boiler and the use of a building as cover, any lost heat was not thought to be as serious. An uncovered boiler would also be much easier to maintain as it allowed ready access to the safety valves and the boiler shell could be inspected quickly.

The heat from the burning coal was drawn under the full length of the boiler shell, the gases were then diverted into two side flues which extended forwards to the fire door end. The two flues then converged into one, allowing the gases to be drawn through the central flue and into the chimney stack and finally discharged into the outside atmosphere.

The plan view drawing dated 29th July 1788, showed that the brickwork of the boiler had a rounded front. This was probably done to improve the flow of hot gases converging at the front of the boiler before entering the central flue.

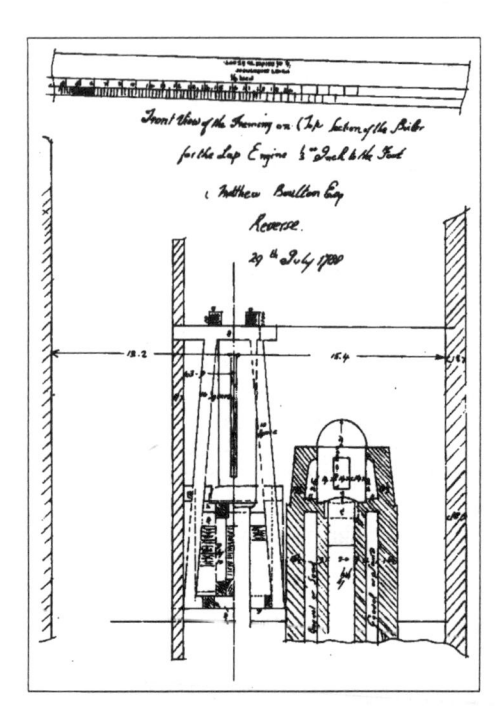

*Figure 46. Matthew Boulton's drawing showing the end view of the Lap Engine and boiler.*

This feature seems distinctive to the Lap Engine as all the subsequent installation drawings show that the surrounding brickwork of boilers was square. As Dickinson and Jenkins mention in their book, '*James Watt and the Steam Engine*', this brickwork must have evolved over many years. An interesting note was written in 1799 and is reproduced below: whether this applies to the Lap Engine is left to speculation!

**The Directions for Erecting Steam Engines.**
   'In building the boiler setting, lime mortar should be used only towards the outside; the parts exposed to fire or flame should be laid in a mortar composed of loam, a sand and clay. Long pieces of rolled iron should be laid in the brickwork to prevent it from splitting and pieces of old cart tyre or iron bars placed under the boiler between it and the bricks to prevent it from being burnt out.'
   No grate sizes for the Lap Engine's boiler have been found, but the usual practice was to apply direct heat on to one-third of the boiler's length. So using this as a brief guide, along with other known constructional details, a grate size 26 inches deep, 36 inches wide and 24 inches long, would not be too different from the grate which was used on the Lap Engine in 1788.

## THE BOULTON AND WATT LAP ENGINE OF 1788

consumed 10½ lbs of best Wednesbury coal per hour for each horsepower. With the Lap Engine rated at 13¾ ihp, it can be calculated that the boiler supplying steam to this engine consumed 145 lbs of coal per hour.

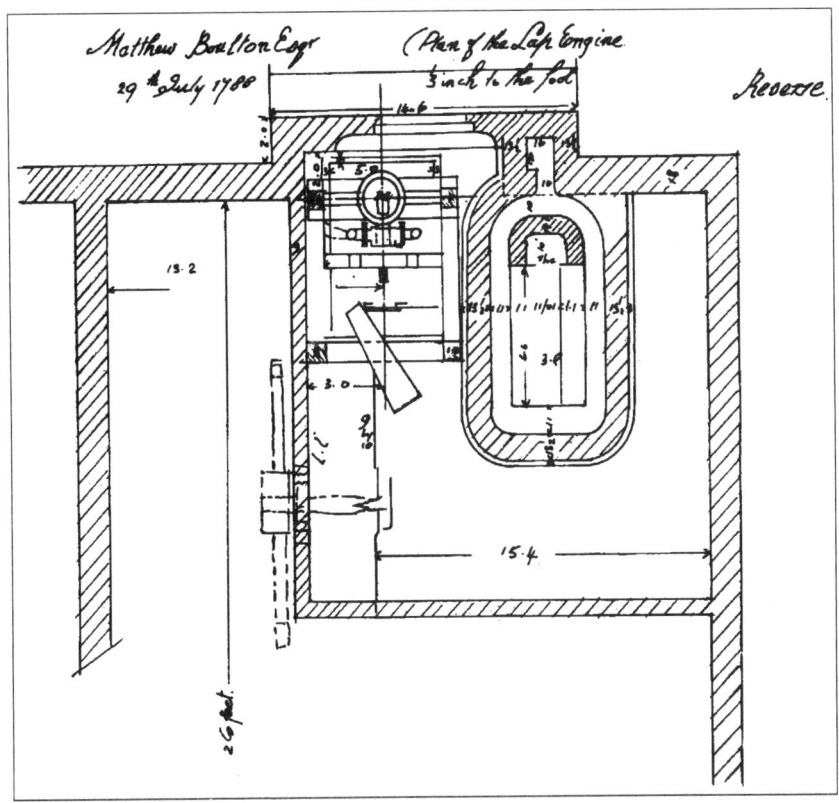

*Figure 47. Plan view of the boiler installation signed by Matthew Boulton on the 29$^{th}$ July 1788.*

### The Construction of the Boiler Shell

The earliest boilers used to power the Watt engines were made of copper, a metal which was malleable, so enabling the intricate shapes to be easily formed. Copper had been used for many years in the brewery trade for vessels and so it could have been considered a natural choice for boilers. However, the use of copper created problems, any accumulation of sediment inside the boiler would have formed hot spots which were liable to burn through to the shell of the boiler.

# THE BOULTON AND WATT LAP ENGINE OF 1788

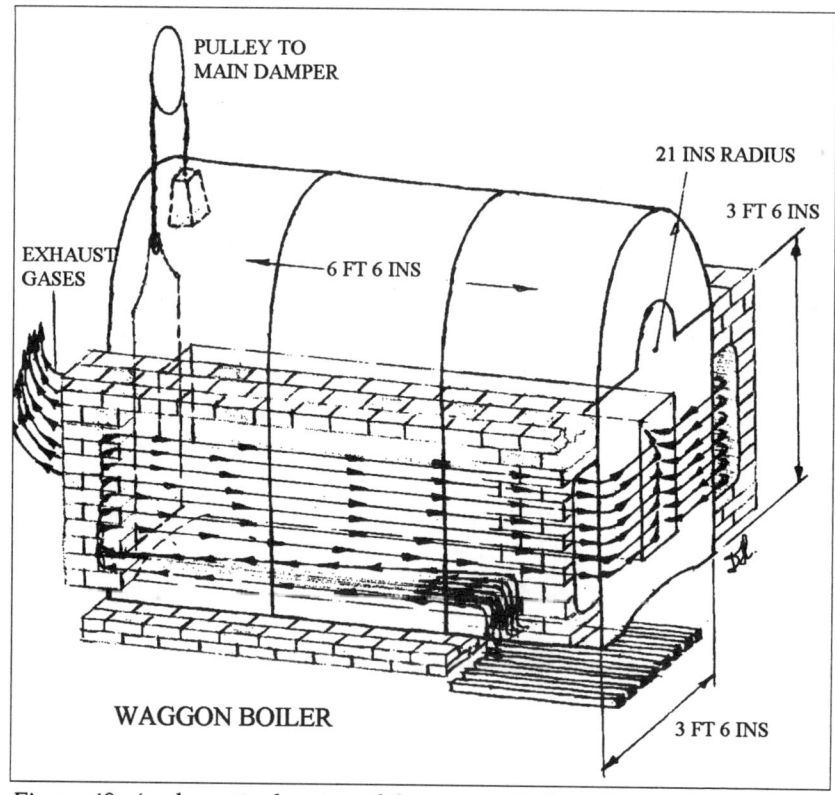

*Figure 48. A schematic drawing of the waggon boiler, showing how the exhaust gases were drawn into the chimney stack.*

With the improvement in manufacturing techniques for boilers the early 1780's saw a change to wrought iron instead of copper for their construction. However, in the early days of the Soho Manufactory, good wrought-iron plates were difficult to produce. Wrought iron billets would be forged into the required flat sheets, but the results were dependant on the type of wrought-iron used. Insurmountable difficulties were found with the use of English wrought iron billets because of their brittle nature, and a much greater success was assured by using wrought-iron imported from Sweden or Russia.

Sweden and Russia produced a very ductile iron which could be easily forged into the intricate shapes needed for boiler production. Because of this, in the early years of the 1780's 'Russian slabs' as they came to be known, were widely used. The *Oxford English Dictionary* states that 'Russian sheet iron' is a ductile sheet iron made in Russia and having a 'smooth glossy surface of purplish colour, sometimes mottled'. All the evidence suggests that the Lap Engine's

## THE BOULTON AND WATT LAP ENGINE OF 1788

boiler plates were made to their final size and thickness at Burton-on-Trent in Staffordshire.

According to C.C. Owen in his book *'Burton upon Trent, The Development of Industry'* the iron billets, arrived from Sweden or Russia at the port of Hull, before passing upstream to be processed in the Midlands. On arrival at Burton Wharf, the billets were collected on behalf of the Birmingham and South Staffordshire hardware manufacturers.

One of these manufacturers, Nathaniel and Charles Lloyd, owned one of the Burton forges where the wrought iron was processed by hammering the billets into flat sheets using their water driven tilt hammers.

Charles Lloyd proudly stated in their literature that Burton Forge supplied Boulton and Watt of Birmingham with iron plates for engine construction. The company produced many of the plates supplied to the Soho Manufactory from the mid-1780s onwards. Iron plates of ⅜ inch and ¼ inch thick were in common use by 1788. The Lap Engine boiler was small and in all probability plates of ¼ inch thick were used, and were put together by using ⅝ inch diameter rivets.

Once the boiler plates had been made, the difficult task of constructing the water and steam-tight pressure vessels could begin. Reproduced below are the actual installation instructions issued by Boulton and Watt for boiler making.

'In making the boiler you should use rivets between ⅝ and ¾ of an inch in diameter.

In the bottom and sides the heads should be large and placed next to the fire in the boiler top and, the heads should be on the inside. The rivets should be spaced at two inch centres and, should be one inch from the edge of the plate.

'The edges of each plate should be accurately cut both inside and out. It is impossible to make a boiler top truly tight which is done otherwise.

'After riveting the edges of the plates should be thickened up. this, is done by a blunt chisel about a quarter of an inch thick, impelled by a hammer of three or more pounds in weight around the whole of the joints.

'After caulking, the joints should be wetted with a solution of sal-ammoniac and water, or urine could be used, which, by promoting rapid corrosion, greatly helped to make the joints steam and water tight. After the boiler is set in the brickwork, the rivet heads were painted with whiting and linseed oil'.

After reading these instructions, one realises how skilled the craftsmen of the eighteenth century were. They could produce a safe steam tight vessel by using hand methods only. The production of the rivet holes must have been a skilled, and noisy undertaking as they were punched by a hand held punch which was said to have been 'struck with a large hammer'.

# THE BOULTON AND WATT LAP ENGINE OF 1788

*Figure 49. This photograph of the miniature engine shows how the safety valves are positioned on the rounded top of the boiler.*

## Boiler Fittings
With the boiler shell now complete, the attachments such as the safety valves and the water feed could now be added.

## The High Pressure Safety Valve
In some boiler installations before 1788, the pressure safety valves, which were situated near the top of the boiler, exhausted the steam direct into the atmosphere. This could have been very unpleasant for the engine operators had they been near the boiler during a pressure blow-off. The excess steam exhausted direct to the atmosphere and was acceptable only when the boiler and the engine were separated by a solid wall. With the Lap Engine and boiler being in the same building, this practice was clearly not acceptable. However, with the development of stuffing box glands, a much more 'user-friendly' and quieter safety valve could be designed. Figure 50 shows this type of safety valve with an adjustable weight which could be moved along a pivoted lever, to increase or decrease the blow off pressure. The working pressure of the steam generally would not require adjustment during the normal running and was set by the

# THE BOULTON AND WATT LAP ENGINE OF 1788

*Figure 50. A sectional view of the high pressure safety valve, designed to prevent a boiler explosion.*

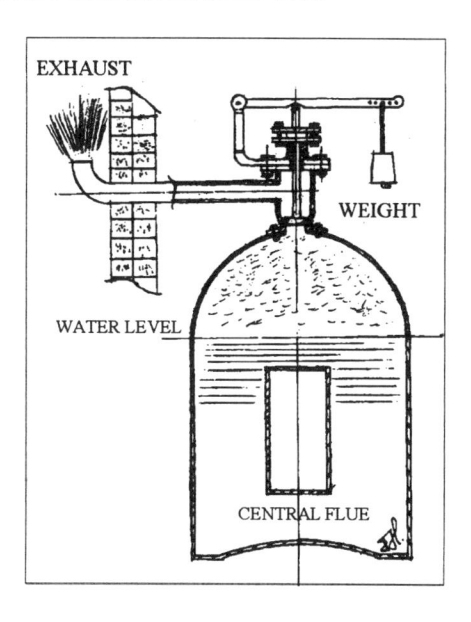

pressure required to completely fill the condenser with steam. This was a necessity for the efficient running of the engine.

The construction of the safety valve was simple, the valve casing contained a coned poppet valve which was held on to its brass seat by an adjustable weight, with the stem of the valve passing through a stuffing box gland. The valve head was made of copper which was riveted onto the wrought iron stem of the valve.

The stuffing box was novel for the 1780's and was designed to enable the Watt engines to work in the new double acting mode. Piston rods were the first to use these glands when they passed through the head of a sealed cylinder.

The stuffing boxes were packed with 'oakum', or hemp soaked in melted tallow. The safety valve released and exhausted the excess steam into the chimney stack through a cast-iron blow-off pipe connected to the receiving chamber of this valve, considerably reducing the earsplitting noise from the excess steam escaping from the boiler.

## The Boiler Feed water Control

The mixture of condensate and fresh water, which had been lifted into the main header tank by means of the engine's secondary pump, flowed by gravity into the boiler feed tank which was installed high above the boiler. This head of water easily overcame the internal steam pressure, without an external pressurised feed. The water required a minimum of seven feet to successfully overcome the internal steam pressure of 2½ pounds per square inch.

## THE BOULTON AND WATT LAP ENGINE OF 1788

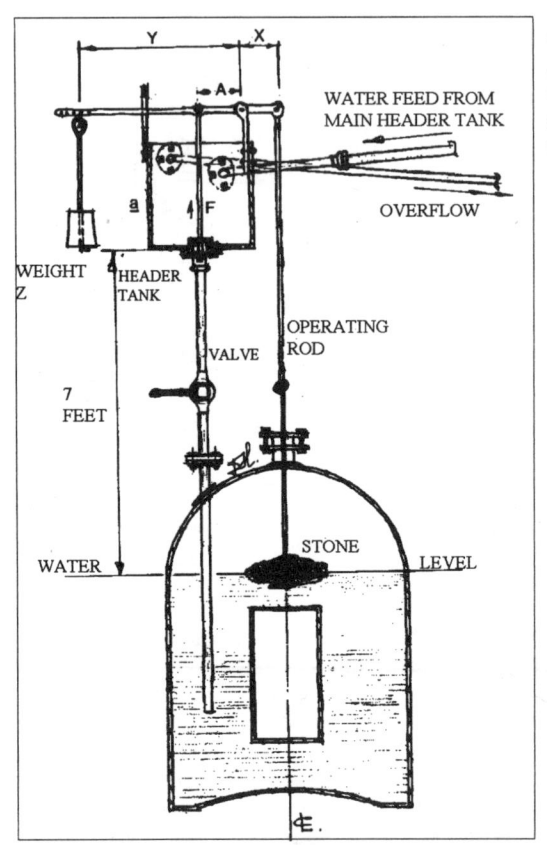

*Figure 51. The mechanism for feeding water into the boiler. Note the stone used as a displacement weight.*

The feed water was retained in the header tank by a valve which was normally held on to its brass seating ring by a cast iron weight. Within the boiler was a balancing stone 'float'*. In order to balance the weight of the stone the cast iron weight was moved along the lever attached to the feed tank.

The stone method was used because corrosion from the boiling water would affect the weight of a metal 'float' which could result in a variable water level. When the boiler water level fell due to evaporation by the production of steam, the stone also lowered, so moving the rod and lever which lifted the cone valve off its seating. The water then flowed in to refill the boiler by way of a vertical standpipe. Positioned midway along this standpipe was a cut off valve which was used to regulate the flow of water entering the boiler. A careful study of Figure 51 will show this very unusual method of control for this filling mechanism.

The stone method of controlling the water level does not seem to have been widely used. Engines contemporary to the Lap Engine, had copper floats.
* *Archimedes Principle was used.*

# THE BOULTON AND WATT LAP ENGINE OF 1788

The development of the stone method of feed water control probably evolved through the need for an extremely accurate method of preventing the water level dropping below the top of the central flue on these small boilers. If the flues were not completely surrounded by water, the hot gases would soon have caused damage and reduced the life of the boiler.

The change to float control, coinciding as it did with the adoption of larger boilers without central flues, might explain the contemporary thinking. A metal float was then quite acceptable as the water level in these boilers was not such a critical factor as it had been previously. So as to enable a consistent head of water in the feed water tank, the supply always had to exceed the demand and, the excess of water then flowed back to the main condensing tank by gravity.

The constructional details of this feeder system were very straightforward: the tank was made of cast iron with a cone valve of copper riveted onto a wrought iron stem. The operating levers were made of wrought iron and cast iron would have been used for the feed and overflow pipes.

## The Main Steam Pipe and Steam Pressure Gauge

The main steam delivery, sometimes referred to as the 'eduction pipe' which supplied the condenser with steam, was positioned at the very top of the boiler. The boilers at this time, working at low pressure, had a turbulent boiling surface and if the main steam delivery pipes were not positioned correctly, an excessive amount of water would be transported along with the steam into the condenser. To reduce this effect to a minimum the 5 inch in diameter steam pipe ascended vertically for the first 24 inches before sweeping into a gentle bend. The pipe was then directed to the top nozzle of the double-acting cylinder. With the eduction pipe positioned in this way any water drawn along with the steam returned to the boiler by gravity, shown in Figure 52.

Positioned midway along the eduction pipe was a steam gauge, the position being determined so it could be easily read by the engineer when he was standing on the operating floor. The gauge consisted of a 'U' tube which contained mercury. Attached to the open end of this tube was a brass scale, which was calibrated in inches of mercury. Suspended by the mercury, the top of a wooden rod was cut to coincide with 'O' on the scale. With the steam at its correct working pressure, this wooden pointer would have read approximately five inches of mercury.

## The Additional Safety Valve

This interesting valve was used to prevent the boiler sides from collapsing by the formation of a vacuum and is shown in Figure 53 on page 73.
After a days work the steam in the boiler would condense, but on condensing a vacuum would form with the possibility of drawing the boiler sides together. If this condition was allowed to arise, the boiler shell would collapse. A simple solution was devised to prevent this, by the use of an additional safety valve. Although the working pressure of the steam was more than adequate to hold this

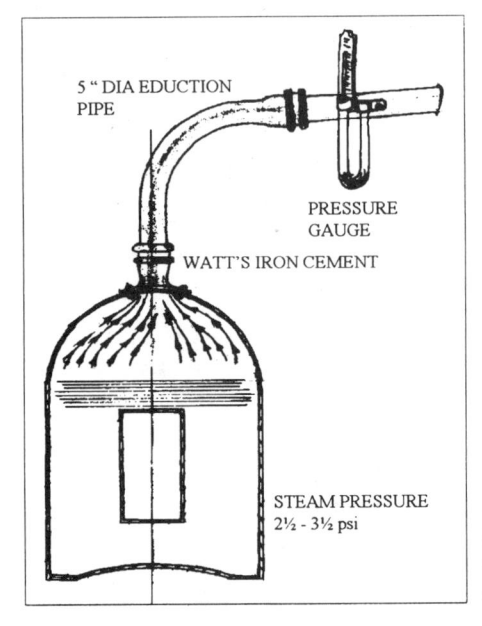

*Figure 52. The steam delivery pipe to the cylinder and into the condenser. Note the use of iron cement.*

valve on to its seating face by the additional balance weight (a) when the engine had finished a working day and the boiler was allowed to cool, the steam within the boiler would condense and the vacuum created would have drawn this valve from its seating face. This would have equalised the pressure inside the boiler to the pressure of the atmosphere and would have prevented any damage to the boiler's wrought iron plates. The cover with this safety valve attached was periodically removed for inspection and the internal cleaning of the boiler. The opening into the boiler was just large enough for a man to squeeze through - so the cleaning of the inside of a boiler must have been a formidable task, said usually to have been carried out by a young boy.

**The Water Level Measurement**

Because of the severe damage which could be caused if the water level in the boiler were to drop below the level of the central flue, a completely reliable method had to be found to keep a check on the level. To provide a backup for the automatic filling system, shown in Figure 51, a physical check was thought to be necessary and, to deal with this problem, two manually operated taps were connected to the stand pipes through holes in the boiler shell. The ends of these pipes had a mean length to coincide with the working water level. When one pipe was shortened by 4 to 6 inches, the lower end of this pipe would then be above the ideal water level and the other pipe would be below.

The engineman would then routinely check the water level by operating these

# THE BOULTON AND WATT LAP ENGINE OF 1788

*Figure 53. A safety valve which prevented a vacuum from forming inside the boiler when the water cooled overnight.*

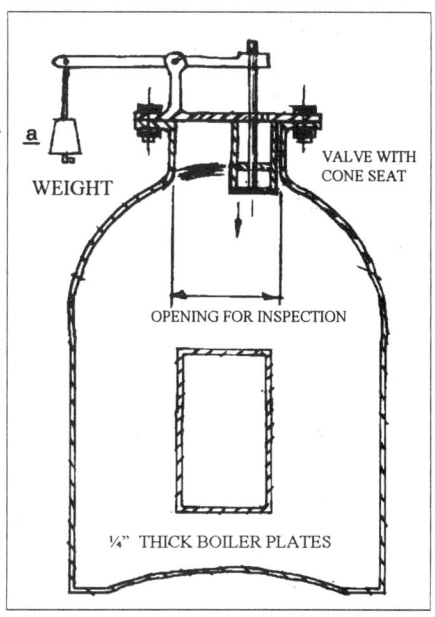

taps. By opening the high-level tap, steam would issue; then by opening the lower tap, boiling water would emerge and, thus the engineman would be assured that the water level was correct between the two pipes inside the boiler. When steam issued from both taps, the indication would be that the water level was too low and some urgent adjustments to the feed water flow control and its associated balance weight would then be needed. The taps were made of brass, with cone-shaped spools. It goes almost without saying that the outlets were pointed away from the operating position.

On the left side of the firedoor was a large drain tap which was lever operated. This was used when the water in the boiler needed to be completely drained off, for example in the case of a breakdown or for regular maintenance.

## The Main Flue Damper

The draught on the fire was controlled by the raising and the lowering of a counterbalanced slide located between the end of the boiler and the chimney stack. After passing the damper, the waste gas was drawn out to the atmosphere. The vertical opening of this damper was governed by the loading of the engine. A careful watch had to be kept on the steam gauge and when the pointer dropped the damper had to be opened and vice versa.

When the fireman was recharging a good fire, wet coals were used for economy because this gave the coal time to fuse together without being immediately drawn into the side ducts and the central flue.

## THE BOULTON AND WATT LAP ENGINE OF 1788

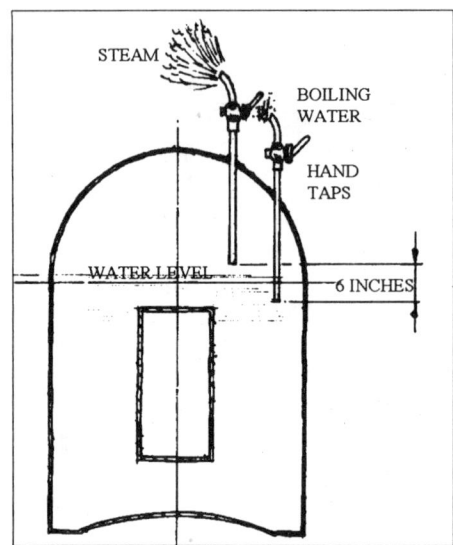

*Figure 54.*
*Water level measurement. The ideal working level is when steam issues from the high level tap and boiling water from the lower tap.*

At the base of the chimney stack on the outside of the building was a small door which was used to regulate the draught on the fire when it was being rekindled. The door was large enough to enable the fallen soot to be cleaned from the chimney and to maintain the maximum draught of air needed for the fire. On an early engraving c 1746, the size of this door was shown to suit the size of the average thirteen-year-old boy! The chimney stack rose to a total height of almost forty feet.

# CHAPTER 6
# THE FORMATION OF A VACUUM AND CYLINDER OPERATION

**The Construction and Operation of the Condensing System.**
How James Watt formulated his ideas on creating a vacuum away from the actual powering cylinder has been described in the introduction to the Smethwick Engine, in my first book. From Watt's originally inspired thoughts of 1765, the separate water-cooled condenser had possibly reached its zenith with the system which was fitted to the Lap Engine in 1788 and the condensing technique was not greatly improved after this date. This design of condenser continued to be made and fitted to Boulton and Watt atmospheric engines until the early years of the nineteenth century.

Engines using the weight of the atmosphere for their power were quickly superseded after the year 1800 by engines designed to work on 'strong steam', which is steam raised to a high pressure and a subsequent high temperature. The elastic property of steam could now be used (these high pressure engines were first developed by Richard Trevithick 1771-1831 and an engine made in 1804 to drive a dye house at Lambeth will be described later). The year 1800 was also the time when James Watt's extended patent finally expired. With the expiry of this extended patent engineers worldwide now had a free hand on steam engine design.

The patent first taken out by James Watt in 1769 covered the following improvements to the operation of the steam engine. One, that the operating cylinder of the engine must always remain hot. Two, that the vacuum should be created in a separate water-cooled vessel Three, that pumps operated by the engine should remove the accumulated water from the condenser. Four, that the expansive force of steam was to be used on the top of the piston. Five, covers the use of steam in a turbine type apparatus. Six, to vary the degree of condensation of the steam.

However the final item which Watt covered by this patent really did halt technological progress as other engineers could not compromise their designs without infringing this method of operation. James Watt actually had patented a free moving piston lubricated within a sealed cylinder by substances other than water.

The patent reads ' Lastly, instead of using water to render the piston or other parts of the engines air and steam tight, I employ oils, wax, resinous bodies, fat of animals, quicksilver and other metals, in their fluid state.'

**Details of the Condensing System of the Lap Engine
and how it operated in 1788**
The steam entered the water-cooled condenser by the way of the eduction pipe (a) which was fitted into a socket cast onto the lid of the condenser, seen in Figure 55. This was then sealed by using Watt's 'iron cement'. The condenser

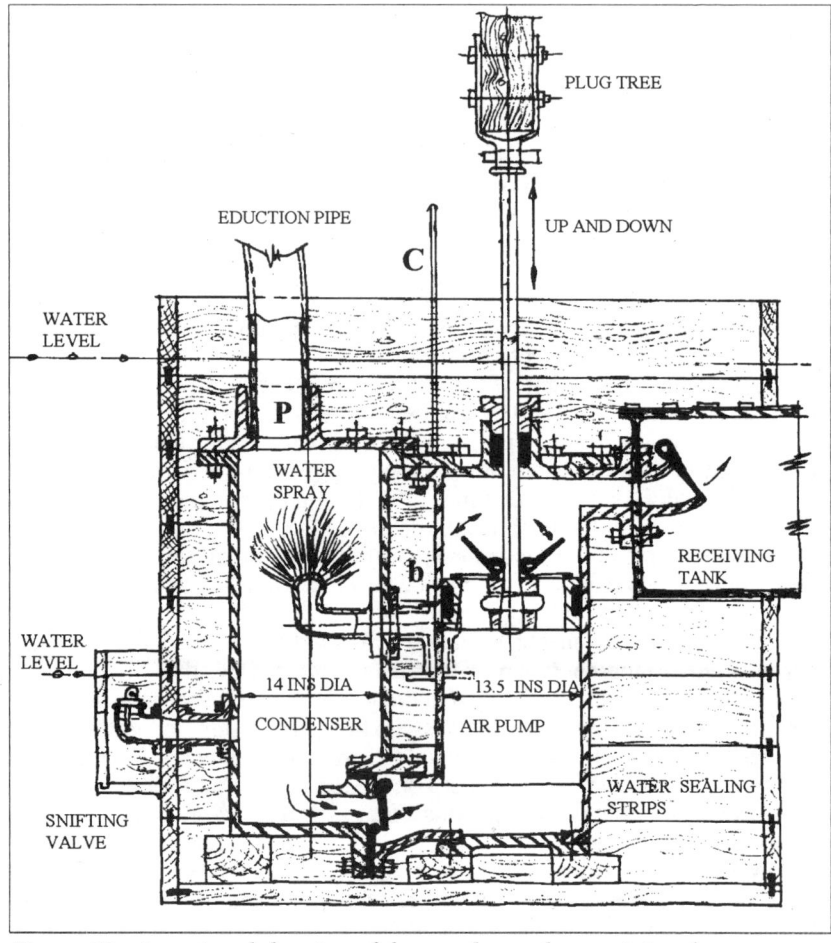

## THE FORMATION OF A VACUUM AND CYLINDER OPERATION

*Figure 55. A sectional drawing of the wooden tank containing the condenser, showing how the components were assembled.*

was a cast-iron closed vessel, the volume of which was equal to half the volume of the main double acting cylinder.

No contemporary relationship for this ratio could be found, but a careful study does reveal that this ratio of two to one was used for all double acting rotative engines. The pumping engines, working on a much slower cycle, show that a ratio of four to one was in use.

The condenser was only a steam receiver which had no moving parts of its own. It was used as cast straight from the foundry without any machining. The end flanges were also used as cast and the seal was then made permanent against

## THE FORMATION OF A VACUUM AND CYLINDER OPERATION

the lid by some quite thick pasteboard gaskets.

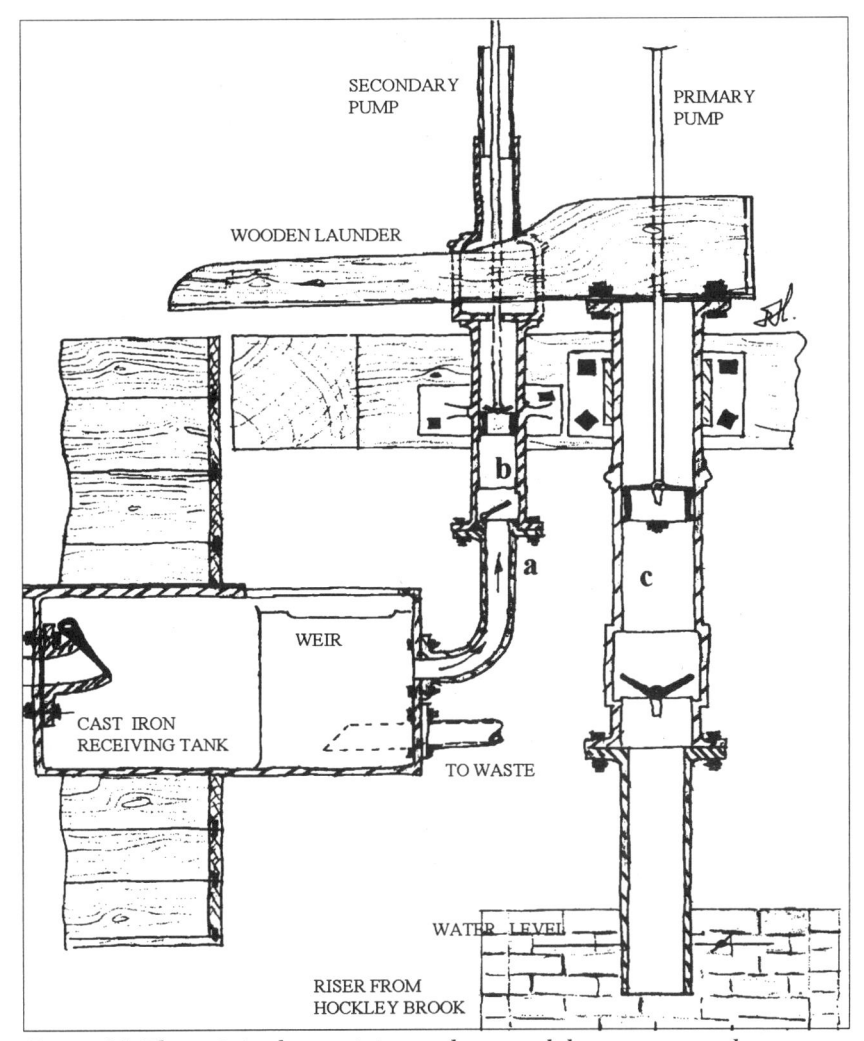

*Figure 56. The weir in the receiving tank ensured that water was always available for the secondary pump. The water which flowed over the weir drained away to waste. View this drawing in conjunction with Figure 55.*

# THE FORMATION OF A VACUUM AND CYLINDER OPERATION

## The Assemblage of the Condensing Components

About halfway down the side the water spraying fountain entered the condenser and this was connected to, and controlled by, the injection valve which was positioned on the condenser's outer wall. The 'fountain' spray itself was created by the inherent water pressure in the condenser tank.

The fountain of cold water condensed the steam within the condenser thus creating a powering vacuum.

The injection valve (b), seen partially obscured by the air pump cylinder in Figure 55, was controlled by a hand-operated lever which revolved the spindle (c) of the valve. The injection valve was made of cast-iron, with a brass conical spool. Positioned about 9 inches above the bottom of the condensing cylinder was a snifting valve, which was submerged in its own small wooden tank attached to the outside of the main cooling tank.

The snifting valve was a conical brass valve which sealed into a brass seating, and was held into position by the force of gravity. This valve allowed the condenser to be completely filled with steam, the build up of air escaping through this valve into the atmosphere. After the steam had condensed the valve retained the vacuum to draw the piston through its powering stroke.

The condensed water, which amounted to ten gallons per horsepower per hour, combined with the water from the spray and was removed from the condenser by the air pump. On the ascending stroke of the pump the water was lifted into the receiving tank, which was positioned inside the main tank just below the working water level.

With the use of the air pump we see the first piece of precision engineering on the Lap Engine. This air pump had a finely bored surface which was produced by John Wilkinson's water-powered boring mill first developed in 1774.

The air pump cylinder had a bore of 13½ inches and the piston made a stroke of 24 inches. The piston was operated by an extension to the lower end of the plug tree, which was lifted up and down by a parallel motion attached to the main oscillating beam of the engine. When the engine was running normally, the condenser and the air pumps contained water to a depth of approximately 9 inches, this accumulated water was only removed when this level was exceeded. The piston of the air pump was fitted with pivoting flaps (or bucket valves) which only allowed water to flow one way, ie. to be retained above the piston and lifted to flow by gravity into the receiving tank.

The complete condensing assemblage, ie. the condenser, the air pump and the receiving tank, were positioned inside the main wooden cold water tank. The condenser and the air pump were clamped into their working position by square headed nuts and bolts passing through to the underside of the tank.

When the condensing system used on these engines was being put together, all the joints were made steam and water tight by the use of thick gaskets made of pasteboard. John Farey quoted 'such as is used for the bounds of books.'

# THE FORMATION OF A VACUUM AND CYLINDER OPERATION

John Farey, in his book, ' *A Treatise on the Steam Engine 1827'.* described how these gaskets were made. 'Soak these pasteboards in warm water until they become quite soft then lay them upon boards to dry, and when quite dry put them into a flat pan with a quantity of drying linseed oil, warm the oil until the pasteboard ceases to emit any bubbles of air'.'With these pasteboard gaskets made, the assembly would be in the normal manner'.

### The Construction of the Cooling Tank

The condensing tank was made of wood 3 inches thick and the recommended timbers were best Danzig deal planks (now renamed Gdansk, Poland). If this timber was not obtainable, good red deal or oak was used. Whatever kind of wood was selected, great care had to be taken not to use sapwood, otherwise the tank or cistern would not last.

In order to hold the planks together, ¾ inch square rods were used, these rods connecting the top to the bottom, passed through holes drilled in each plank. Before clamping the planks together, the edges were covered with strips of coarse flannel, soaked with a mixture of tar and tallow in equal parts, and the tar and tallow was applied when hot. If flannel was not available, bull rushes could also be used.

The water level inside the condensing tank was controlled by a weir, with the water at the correct level the water flowed over this weir and then drained away to waste. A large brass valve was situated at the bottom of the tank and was used when the tank had to be drained off for any maintenance or when a breakdown occurred.

A detachable base was fitted to the air pump cylinder to enable the casting to be finished on John Wilkinson's cylinder boring machine. These water powered mills were unable to machine a cylinder with a 'blind end', as the cutting head had to be mounted between the centre support bearings on the boring machine.

### The Cold Water Pump and the Hot Water Pump.
### ( Also referred to as the Primary and Secondary Pumps)

The hot water from the condensing cylinder of the Lap Engine was lifted into the cast iron receiving tank by the air pump positioned adjacently to the condenser. From the receiving tank, the water was again lifted to the engine's header tank by the use of a conventional bucket lift pump, ie secondary pump (a) shown on Figure 56.The water could then flow by gravity into the header tank positioned 7 feet above the dome of the boiler.

### The Cast Iron Receiving Tank

To make sure that this tank always had water available for the hot water or secondary pump, a weir divided the tank into two separate compartments. Only excess water passed over the weir from the receiving tank before flowing away to waste. Unfortunately there are no constructional details to be found for the watertight joint where the cast iron receiving tank projected through the side

## THE FORMATION OF A VACUUM AND CYLINDER OPERATION

wall of the wooden cooling tank, but the joint was probably sealed by flannel soaked in tar and tallow. The fit would have been quite critical to avoid leakage as the two tanks expanded and contracted over the working day, and maintaining the watertight joints must have required considerable care.

### The Hot Water Pump.

The design of the 'bucket' pumps followed very closely the principle of the old village pump which would have been used in rural communities for domestic water. The water from the receiving tank was drawn up and through the flap (a) Figure 56 by the ascending piston or bucket (b). On the descent the water passed through the one-way flaps fitted into the piston, ready to be lifted on the next ascending stroke into the engine's header tank.

The piston was most probably made of leather with flaps of cast brass; and worked inside the cast-iron cylinder. The only deviation from a village pump practice was that each pump had to be designed very carefully to determine the pumping capacity for each stroke of the engine.

The hot water pump had to deliver a much greater quantity of water into the boiler feed water tank, or cistern, than the amount of water which evaporated from the boiler in the same time. The surplus then flowed away to waste. Because the feeding of water to the boiler could be accidentally interrupted and, the water level in the boiler could then fall to a dangerously low level when this deficiency was discovered the boiler had to be refilled as quickly as possible.

If the hot water pump could not deliver more water than was required to keep the boiler at its ideal operating level, the pump would be incapable of making up this deficiency. The greater the amount of water this pump was able to deliver, the quicker the water level in the boiler could be restored.

### How the Diameter of the Hot Water Pump was Calculated

Watts' directions - 'Divide the square of the diameter of the cylinder in inches by 240, multiply the quotient by the radius in inches, from the main joint of the piston rod. Divide the product by the radius, in inches of the joint for the hot water pump, the square root of the quotient is the proper diameter for the hot water pump in inches'. This summary of how to calculate the diameter of the water pump has been taken from John Farey's book, *'A Treatise on the Steam Engine'*.

### The Primary or Cold Water Pump

The action of the primary or cold water pump was identical to that of the hot water pump already described. The main difference was that the cold water pump had an even larger pumping capacity than that of the hot water pump in order to maintain the cooling tank at a constant low temperature.

The arrangement of this pump is illustrated in Figure 56 (c). There is no material evidence of the existence of this pump nor of the hot water pump: All the details stated here were taken from contemporary drawings.

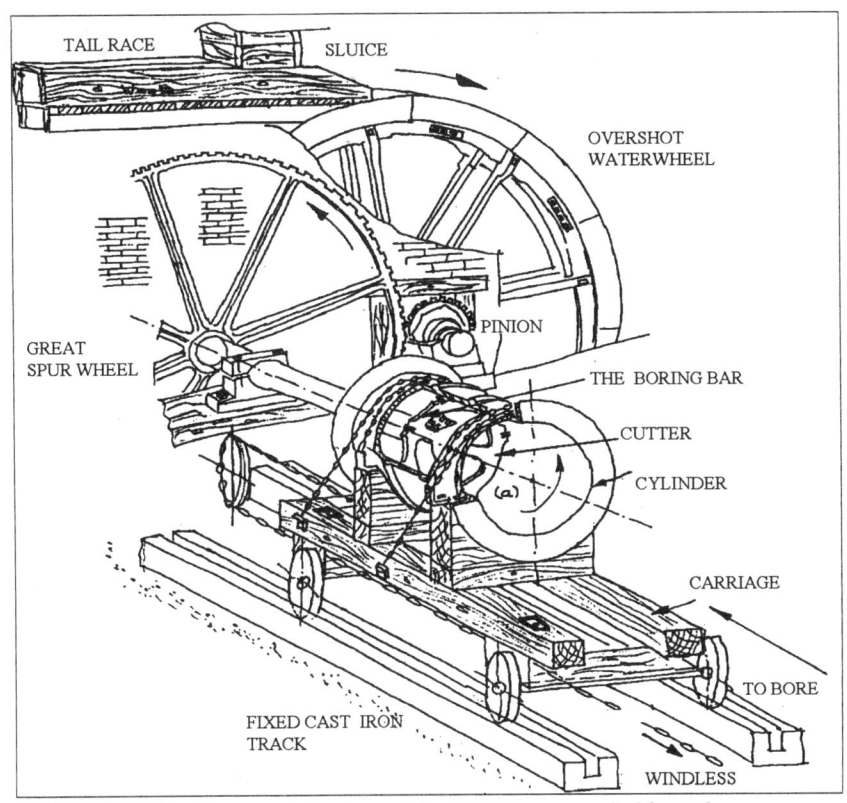

# THE FORMATION OF A VACUUM AND CYLINDER OPERATION

*Figure 57. John Smeaton's boring mill of 1769 superseded by John Wilkinson's mill of 1774 which was shown in Figure 23. page 34.*

The primary water which was needed to run the Lap Engine was pumped from Hockley Brook which flowed out of Soho Pool by the Manufactory.

The pumping capacity of these cold water pumps equalled 1/48 of the volume of the main steam cylinder. These pumps had an efficiency of only 40 per cent, because the water always escaped through the flaps. When the engine was working, every horsepower required forty cubic feet of water per hour. From this, the water consumption of the Lap Engine can be calculated as 14 ihp which would require 550 cubic feet or about 3,400 gallons of water per hour.

The pumps fitted to the Lap Engine were operated by vertical rods connected to the engine's main oscillating beam.

Most references to the engine show that the delivery launder from the primary pump into the condensing tank was made from cast iron, though a careful study of the drawings (Figures 46 and 47) will reveal that this part was originally made of wood when the engine was first assembled in 1788.

## The Development of the Double-Acting Cylinder

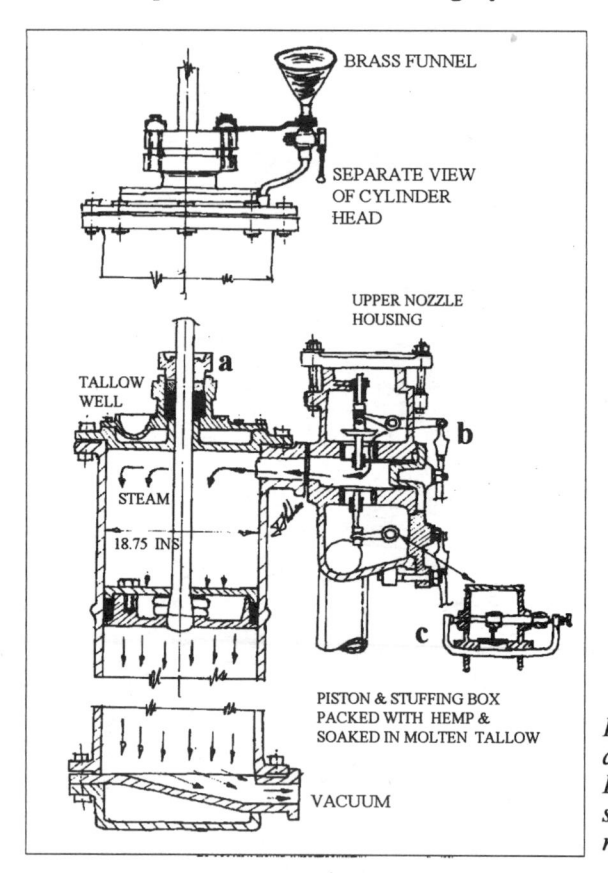

*Figure 58. The double-acting cylinder of the Lap Engine showing in section the piston and nozzle housings.*

The two methods used to machine the internal surface of a cylinder bore have been briefly described, with the description leading into the manufacture and installation of the Smethwick engine, in 1779, (described in my first book).

It was the improvements between John Smeaton's boring mills and a new cylinder boring machine designed by John Wilkinson that played a major part in the rapid development of the rotative engine in the latter part of the eighteenth century.

John Smeaton (1724-1792) was a civil engineer who made many improvements to the Newcomen engines. However, these improvements did not upgrade the engine's poor thermal efficiency, they only resulted in better proportioning and construction of the operational parts. With this object in mind, Smeaton designed an improved boring mill which could be used for the machining of steam engine cylinders (Figure 57). This mill was driven through

*Figure 59. The cylinder showing all the valve operating levers. Also shown is a part of the wooden support structure of the engine.*

a reduction gear by an overshot waterwheel. The cutting head on this boring machine had six tool bits made of cementation steel (high carbon steel) and attached to a strong 'axle' driven by the great spur wheel.

The Smeaton boring mills, although they contained no great innovation, were a great improvement over earlier methods of cylinder production. However the cantilever method of securing the boring bar had a tendency for deflection, with the result that the cylinder bore could become slightly oval in shape. In an endeavour to overcome this inaccuracy, the cylinder had to be halved or quartered, ie. the cylinder had to be turned through 90 degrees or 180 degrees and traversed again over the cutting head with the same applied cut. Even with all this care the resulting cylinders were still not accurate enough for the requirements of James Watt's double acting cylinders.

**John Wilkinson's Boring Machine**

In 1774 the eventual breakthrough came with a cylinder boring machine which was designed by John Wilkinson, Figure 23. Wilkinson developed an accurate method of guiding the cutting head through a fixed and stationary workpiece. The cutting head held six tool bits and was traversed by means of a rack and pinion along an accurately machined bar.

The boring bar had a longitudinal slot extending along its full length and through this slot the necessary attachment was made between the internal pusher bar and the cutting head. The boring bar in this mill was supported by bearings at each end and the bar was large in diameter resulting in very little deflection of the boring or cutting head.

The development of Wilkinson's boring machine enabled steam engine design to progress rapidly and the introduction of this machine was probably just as important an event to production engineering as James Watt's double-acting cylinder was to the rotative steam engine. Perhaps the ten years of frustration and delay suffered by James Watt in gaining acceptance for his separate condenser was not all wasted because, without the invention of John Wilkinson's boring machine, the double-acting cylinder could not have been made.

For the first time pistons could be sealed in their cylinders using only lubricated hemp without the addition of water as was previously used in all Newcomen type engines. This allowed the piston to be powered in both directions, ie. double-acting, and the first rotative engine to use this double-acting principle was built in 1783 and put to work at the Boulton and Watt Soho Manufactory.

### The Double-Acting Cylinder of the Lap Engine

With cylinder bores now accurate enough for the hemp packing around the piston to be the only sealing agent required, this brought new problems. Firstly, how do you seal a piston rod passing through a cylinder top? Secondly, how do you lubricate the hemp packing around the piston which is powered by a vacuum inside a closed vessel?

Fortunately, both these problems had a simple solution - the introduction of the stuffing box which was first used on the Smethwick Engine in 1779. The seal which was fitted to the Lap Engine is shown in Figure 58 (a). This stuffing box was simply a circular cavity, packed with soft rope yarn soaked in melted tallow, before all being clamped into position by a wooden spacer and a brass gland. The wrought iron piston rod had to be very finely finished and dimensionally accurate if the newly developed stuffing box was to be an effective seal for vacuum and steam for a long period of time.

The smooth finish on the piston rod resulted from the rod being turned on a specially designed lathe, installed at the Soho Manufactory in 1777.

The solution to the lubrication problem inside the cylinder was a very ingenious one. Cast into the cylinder head was a well into which a blend of tallow and olive oil was placed. The heat from the incoming steam heated the cylinder head which in turn melted the 'blend' into a manageable lubrication fluid. In this molten state, the tallow was ladled from this well into the brass funnel which was attached onto a stand pipe leading into the cylinder. With the piston at the bottom of its stroke the engine man could quickly operate the *on-off* valve and the vacuum created would then draw the tallow and olive oil out of the

## THE FORMATION OF A VACUUM AND CYLINDER OPERATION

funnel and into the cylinder. Operated in this way, the piston and cylinder were efficiently lubricated. Prompt action was required so that the valve could be quickly returned into the *off* position before the next stroke of the piston.

When glancing back to the air pump previously described, we see that the piston of the air pump was packed with hemp in exactly the same way as the main piston of the Lap Engine.

The length of the main powering cylinder casting was 62 inches and, after a working life of about seventy years, the bore measured 18¾ inches. This is thought provoking because all contemporary references to engines constructed c1788 state the diameter of the cylinders in whole numbers and the reference to the Lap Engine is seen as 18 inches. Where did this additional ¾ inch come from? Perhaps from numerous re-bores during the engine's long working life?

The Lap Engine's theoretical piston stroke of 48.98 inches at first seems odd but this stroke was, of course, equal to the diameter of the orbit traced out by the planet gear's movement around the sun gear. Both the sun and the planet gear have equal pitch circle diameters and these are 24.49 inches. From this we can deduce that the planet gear's orbit was double this measurement and equalled 48.98 inches which was the stroke of the main cylinder.

The sun and planet gearing itself will be described in detail later, see page 101 Figure 58 also shows that the cylinder had a sloping base leading into the eduction pipe. This was designed to prevent any build up of water under the piston. Could this slope also have helped the flow of the vacuum to the condenser or are we thinking of modern gas flow techniques which perhaps did not apply to this early engine?

When the final assembly of the cylinder took place each of the end joints was made vacuum and steam tight by the use of pasteboard gaskets, all held in position with six ¾ inch diameter square-headed nuts and bolts. The whole assembly was mounted on to two strong oak beams and secured by the use of heavy clamps. The double-acting cylinder made for the Lap Engine in miniature is shown in Figure 59.

**Nozzle or Valve Housings**

The nozzle housings were of a complex nature and must have created many pattern making and casting problems. The two housings and the interconnecting eduction system, were reproduced as a single casting and a close study of Figure 59, together with the interconnecting eduction system, shows the complicated nature of this casting. It is difficult to represent this fully on paper because of the surrounding valve gear. However, a study of the prototype displayed in the South Kensington Museum will reveal how skilled the iron founders of the eighteenth century really were. The inside of the nozzle housings had to have easy access both for assembly and the setting of the valves.

To facilitate this each housing was fitted with three cast-iron inspection covers (b), each one held being in position by eye bolts and nuts.

The drop valves were of an unusual construction because the copper heads

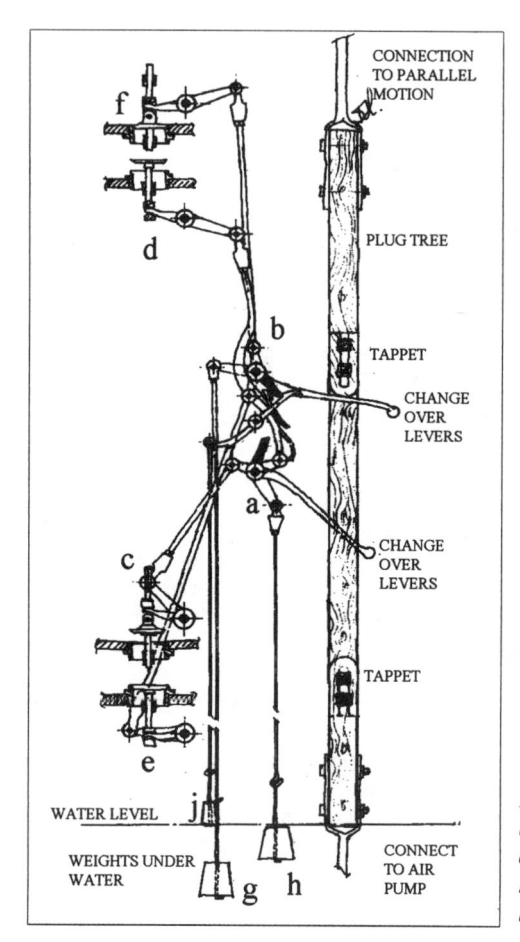

## THE FORMATION OF A VACUUM AND CYLINDER OPERATION

*Figure 60. A view showing how the drop valves were opened and closed by the up and down movement of the wooden plug tree.*

were riveted to wrought-iron stems. The valves made their seal against a brass seat, which was press-fitted into the cast iron casting. Also of special note are the rocker arm retaining clamps, which hold the spindles into their working position. These clamps are in the form of a 'G'-clamp (c). The rocker spindles are held between centres by two 60 degree conical retainers (like turning between centres on a lathe). With this type of restraint the rockers and spindles could easily be assembled from one side of the housing, Figure 59.

**The Operation of the Nozzle Housing** (Valve Chest)

The expression 'nozzle housing' was used in the eighteenth century to describe the chest containing the operating valves of the cylinder. These nozzle housings were operated by what at a first glance looked like a multitude of

## THE FORMATION OF A VACUUM AND CYLINDER OPERATION

levers, pawls, catch-bars, leather straps with buckle adjustments, and weights swinging back and forth, also moving up and down. All of the levers were manoeuvred by tappets attached to the side of the wooden plug-tree on both the up and down stroke of the powering cylinder. However, the movements of these levers are not as complicated as they first appear. On making a careful study of these valve operating levers it all devolves into a rather simple operation.

The great number of components controlling the valves developed from the need to operate two valves in each housing, with one housing being positioned at each end of the powering cylinder. With the cylinder being powered in both directions the levers were simply 'doubled up' and each housing was operated in exactly the same way, alternating on each cycle or stroke of the piston.

The operating levers are pivoted on two horizontal shafts constructed of wrought-iron, Figure 60 (a) and (b). When in the lower position the plug-tree operates the tappet mounted on the shaft (b) and this allows the steam to enter into port (c) by way of a drop valve; the steam then issues from the boiler through the eduction system to the underside of the piston.

At the same time drop valve (d) also opens allowing a direct connection (p) see Figure 55, to the vacuum which had been created inside the condenser. With these two drop valves open the piston was powered to the top of the cylinder.

When the plug-tree had lifted 24 inches vertically drop valves (c) and (d) were closed and drop valves (e) and (f) open, allowing the piston to be powered in the other direction.

The amount of pressurised steam on one side of the piston, and the degree of vacuum on the other side of the piston, was controlled by the adjustment of leather straps fitted with brass buckles. These straps controlled the downward movements of weights (g) and (h). The only adjustment ever needed was the initial setting unless a breakdown occurred.

On the up and down movement of the plug-tree, the changeover levers held the drop valves open until the engagement between the pawl and the catch bar had taken place. The weights (g) and (h) were used to hold their respective pairs of valves open. The weight (j) was only used to hold the catch-bar against a fixed stop, allowing the catch-bar to arrest the angular movement of the pawls on each engine cycle.

The engine on display in the South Kensington Museum has its plug-tree constructed of wrought-iron. In the engine's long working life (1788-1858), a replacement must have been made as the original drawing clearly shows the plug-tree was made of wood in 1788.

When the valve operating mechanism was completed on the model a very interesting fact emerged which never came to light through research. Most anxious to see all these levers in action the model was powered by a small electric motor which ran the engine at 25 oscillations per minute, (the speed of the original engine in 1788). However, at this speed this valve gear would not function correctly, because the suspended weights (g). (h) and (j) became entangled. Consequently, this made the operating levers move in all directions! However, a method of operating the valve gear successfully soon became

## THE FORMATION OF A VACUUM AND CYLINDER OPERATION

apparent. The weights were submerged in a cup of water, as they would have been submerged inside the condensing tank of the engine in 1788.

The water, acting as a damper, controlled the movement of the weights and converted the erratic movement of the valve levers into one of harmonic motion. For demonstration purposes, the model has to work at half the normal speed of the original engine if these valve rods are to function properly. The model is now too delicate to have its condensation tank filled with water.

The engine when it drove the lapping and polishing machines at the Soho Manufactory normally worked at 25 cycles per minute, giving an output speed of 50 revolutions per minute, the sun and planet gears increasing the output speed by a ratio of one to two.

*Figure 61. The valve operating levers to the double-acting cylinder of the original Lap Engine in the Science Museum, London. The plug tree on the left is a (relatively) modern replacement- the original wooden plug tree must have worn out.*

# CHAPTER 7

# STRAIGHT LINE MOVEMENT FOR A PISTON ROD

**Watt's Parallel Motion**

During his long working life James Watt said that he gained the most satisfaction from his invention of converting a radial displacement into a rectilinear or straight line movement. A mechanism to achieve a movement of this type was vitally important to the efficient operation of a beam engine which was operated by a cylinder powered in both directions. This mechanism later became known as Watt's parallel motion. He first thought of this theory in 1784 and in a letter to Matthew Boulton, written in the same year, he told him of this idea, which is quoted in Dickinson and Jenkins, *'James Watt and the Steam Engine'*.

'I have started a new hare, I have got a glimpse of a method of causing a piston rod to move up and down perpendicularly, by only fixing it to a piece of iron upon the beams, without chains, or perpendicular guides, or untowardly frictions, arch-heads or other pieces of clumsiness, by which contrivance if it answers fully to expectations about five feet in the height of the engine house may be saved in engines with eight feet strokes'.

Watt goes on to say, -

'I have only tried it in a slight model yet so cannot build upon it, though I think it a very probable thing to succeed, and one of the most ingenious simple pieces of mechanism I have contrived'.

In order to try out this new concept, James Watt built a large-scale model which used his parallel motion design and this model was so successful that he took out a patent for his idea on the 24th August 1784 in anticipation of putting it into operation on a full size engine.

So that the power from a double acting cylinder can be successfully transmitted to the main oscillating beam, the need for an efficient method of parallel movement is vitally important if the engine is to work successfully without applying any sideways force to the piston rod. On single-action engines, the oscillating main beam was constructed with radiused arch-heads which had been used with great success on Thomas Newcomen's engine of 1712. This method of working, with a chain following the radius of the arch-head, was very satisfactory when the force was applied in tension only. However, with the new double-acting engine design, a new method of connecting the piston rod to the oscillating main beam had to be found. Watt tried many ways to achieve parallel motion and many of these early methods were unsatisfactory because they were inefficient, noisy and many of the ideas reduced the power output of the engine.

Before describing how James Watt achieved his parallel motion, a small synopsis of three other methods used to harness the power of a double acting cylinder is relevant.

# STRAIGHT LINE MOVEMENT FOR A PISTON ROD

## Methods used before James Watt

The most common method of producing a parallel motion was by the use of two wrought-iron chains held in tension.

A single-acting engine developed a force in one direction only and, to transmit this force, a chain was compelled to follow the radius around a large wooden attachment on each end of the main beam. The outer edge of these baulks of timber had a radius which was equal to the distance from the main pivot of the rocking beam to the centre line of the cylinder. The curved timbers were usually fastened to both ends of the rocking beam and they were called 'arch heads' or, on some engines became known as ' horses' heads', because of their shape.

The 'arch head' method again harnessed the power of the newly developed double-acting cylinder but this time, three tension chains were used. Two of the chains were attached in the normal way, ie. one end of each chain was connected to the piston rod, while the other end of the chain was connected to the upper edge of the arch head. These chains could then transfer the downwards powering force of the piston to the main oscillating beam.

The additional power produced by a double-acting cylinder, ie. the upward force from the piston, was then transferred to the main beam by firmly attaching a strong wrought iron extension bar to the piston rod. This extended rod, slightly longer than the working stroke of the cylinder, held a third chain, which was connected from the top of this rod to the lower edge of the arch head. This additional chain would then pull the main rocking beam vertically up and thus transmit the double force from the powering cylinder.

With the three chains correctly adjusted, the double-acting force produced by the newly developed cylinder and piston could be transferred to the main oscillating beam. Before 1784, this method of achieving parallel action to the piston rod had been the most successful until the development of James Watt's method.

The second method used to achieve parallel action on a double-acting cylinder used a rack and sector. The rack was a cast-iron extension securely bolted to the upper end of the piston rod and had teeth formed as a gear along its full length. The length of the rack had to slightly exceed the working stroke of the cylinder, the rack then meshed with gear teeth which were formed on to the outer edge of a cast-iron sector which was securely bolted to the face of the wooden arch head of the main oscillating beam.

The rack and sector method of parallel motion caused great problems because it was very difficult to keep the teeth of the rack in constant mesh with the teeth of the sector. The tangential force produced between the engaging teeth tended to force the two gears apart.

A third method was proposed, but it is not known if this idea was ever used on a beam engine. This method was to fit two cast-iron slides, one positioned on each side of the piston rod. A cast iron block would then have been attached to an extended piston rod. This iron block would then have moved up and down vertically by engaging with the two slides. If this method had ever been used, it would have given true straight line movement to the piston rod. This was the method used a hundred years later in the manufacture of table engines.

The great problem in the 1780's was that there was no accurate method of manufacturing the side guides. Only hand methods of production were available and, in these circumstances, an accurate fit for a sliding block would have been difficult to achieve.

This theoretical method could have been the inspiration which James Watt needed, as the sliding guide principle was not too different from the actual idea of parallel motion which Watt eventually perfected. To make the sliding guide method work, the links (a) shown in Figure 62 would have had to have been used to connect the piston rod to the main oscillating beam of the engine.

### The Parallel Motion as fitted to the Lap Engine

The parallel motion which was fitted to the Lap Engine was, of course, the system described by James Watt in his letter to Matthew Boulton. This system was a workable solution to the problem of the parallel motion and, as a result of this, a much more compact engine could be designed and the engine houses reduced in height. All other forms of parallel motions had needed taller buildings to accommodate the extended piston rods. As stated by James Watt, the saving in height on these buildings would amount to about 5 feet on an engine with a piston stroke of 8 feet

### How James Watt's method worked. ( see Figure 62)

The connection between the piston rod and the main beam was made by means of a lever which, Watt called a great perforated link (a). This link was allowed to pivot both at the connection with the main beam and the piston rod. Two such links were fitted and these were held into their working position by a wrought iron spacer.

The up and down movement of the main beam caused the radius bars (c), which did the regulating to pivot at a fixed point bearing (d). The radial movement created by the beam of the engine is then converted into a lateral displacement at the fulcrum (e). Connected to the fulcrum (e) is the parallel bar (b) which then conveys this lateral displacement to the piston rod pivot (f) and compels the pivot (f) to move in a straight line when the great perforated link pivots at (g).

The lateral displacement of (e) is then equal to the lateral displacement created by the radial movement of the main beam, and when these two conditions coincide, true parallel motion is the result. This method of parallel motion had reached its pinnacle and was so successful that it was never superseded in principle. Watt's inspired design was in continuous use on beam engines until the beginning of the twentieth century. The principle became so synonymous with James Watt that it came to be called Watt's parallel motion.

The most important dimension in this pivoting parallelogram of levers is the length of the regulating radius bar (c), ie. the distance from the centre (d) to centre (e).

## STRAIGHT LINE MOVEMENT FOR A PISTON ROD

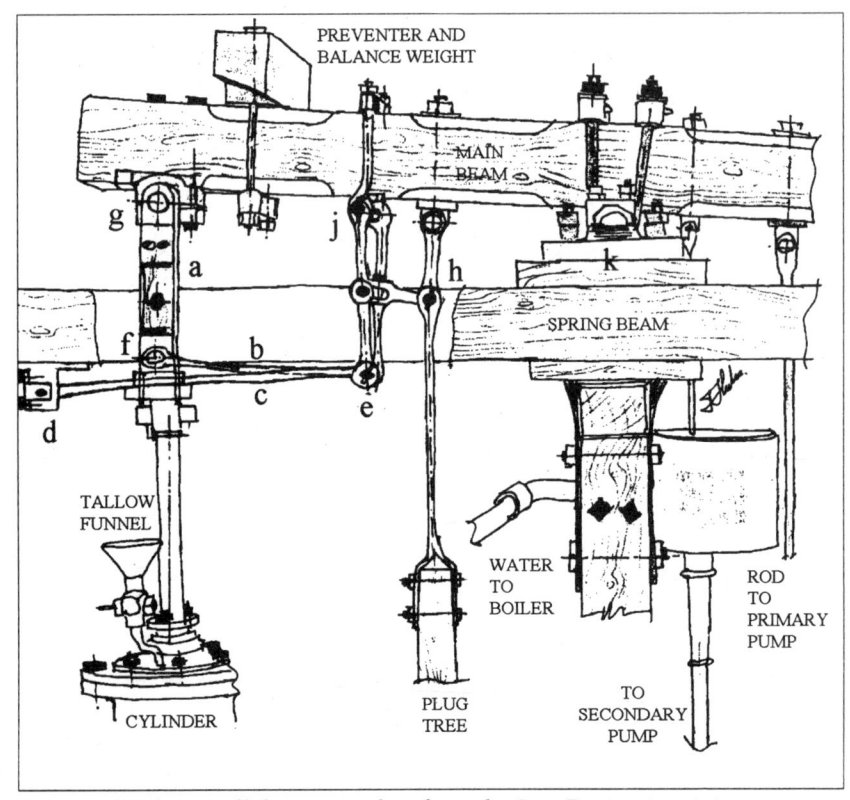

*Figure 62. The parallel motion taken from the Lap Engine in miniature, showing how the radius bars all fit together.*

A simple formula, communicated to John Farey from a Mr Benjamin Hick of Bolton, was as follows: square the distance (k) to (g) (Figure 62) and divide this measurement by (g) to (f) and this should be the theoretical length to the centres of the regulating radius bars (c). All the dimensions were described in inches.

In addition to the parallel motion transmitted to the piston rod on the Lap Engine, there was also an additional parallel motion (h) to aid the vertical movement of the plug tree. This last movement was connected to the midpoint of the link (e). The midpoint connection is required because the plug-tree has only 24 inches of movement compared to about 48 inches for the piston and as such, has half the radial displacement at the piston rod.

# STRAIGHT LINE MOVEMENT FOR A PISTON ROD

**Details of the Parallel Motion as fitted to the Lap Engine**

The connection between the piston rod and the main beam of the engine, was made by two great perforated links (a), the name of which was first used to describe the links of an engine erected in 1785 at the Albion Mill in London. These links consisted of split bearings made from bronze, held at fixed centres by a wooden spacer. This whole assembly was held together by a wrought-iron strap and then finally clamped by the means of the traditional gib and cotter arrangement. On these timber-framed engines, a degree of movement of the joints was necessary and when the alignment had been made, the gibs and cotters could be finally driven into their working positions. These links were fastened to the main oscillating beam by clamping a 1½ inch diameter double ended axle between two iron castings. The whole assembly was then held onto the beam by two strong nuts and bolts.

The two fixed plummer blocks (d) were unusual because the bronze bushes were secured from the side: which allowed a degree of horizontal movement when the engine was being assembled. This was achieved by the use of metal shims. If conventional plummer blocks had been fitted, the fine horizontal adjustment necessary would not have been possible.

All the levers and radius bars used in the parallel motion were forged from wrought iron. On first seeing the engine in the Science Museum at South Kensington, all the component parts making up this great innovation appear to be of very small proportions. The radius rods and the parallel bars are only ¾ inch in diameter. In the early years of the development of the rotative beam engine such small proportions could have led to many rod failures. Many of James Watt's customers were very reluctant to accept these changes, with the result that the arch head method of parallel motion for the piston rod was not completely phased out until after 1792.

**The Condensate Header Tank**

This small cast-iron tank had a capacity of approximately ten gallons and was positioned just below the top of the engine's wooden 'A' frame. The tank acted as a receiver from the secondary or hot water pump and was positioned high enough to enable the water to flow by gravity into the boiler re-feeding tank.

The operating rod to the water feed pump was attached to the main beam of the engine, the rod then passed through both the header tank and the vertical cast iron pipe to operate the pump which was positioned just above the wooden condensing tank.

**Spring Beams**

These 36 feet long beams with a section of 9 inches by 9 inches would have been made from English oak. As the length of these beams was so great, they would have been made in two pieces and joined together with lapped or scarf joints and held with wrought iron straps and bolts.

# STRAIGHT LINE MOVEMENT FOR A PISTON ROD

### The central Bearings supporting the Main Beam.

The central bearing or plummer blocks of the engine were made from cast iron and were fitted with plain, bronze bushes. In order to prevent the main oscillating beam from any sideways movement, these bearings had a very unusual feature: the lower half of the casting had a blind end which restrained the ends of the main central shaft. Arranged in this way the ends of the shaft were the only accurate machining which was necessary. Each plummer block housing was held into its working position by two ¾ inch diameter bolts which passed through the spring beams.

The whole engine was very carefully designed to minimise accurate machining. Many of the component parts were assembled as they had been made by either casting or forging.

### The Main Beam

The main oscillating or rocking beam of the engine, would have been made from the best English oak available and was 15 feet long, 14 inches deep with a width of 12 inches. Beams were accurately cut and very well seasoned. Almost all the beams used at this time by Boulton and Watt were made of oak, although red deal was sometimes used on engines erected in Cornwall.

There were two rules used to determine the size of the main beam of Mr. Watt's Rotative Steam Engine as it was described in John Farey's book, *'A Treatise on the Steam Engine 1827'*.

*To determine the depth,*

Divide the diameter of the cylinder in inches by 1.2. Therefore, in the case of the Lap Engine, 18¾ inches divided by 1.2 equals 15⅝ inches.

*To determine the width,*

Divide the centre distance from the piston rod to the connecting rod centre by 12. Therefore, 159⅔ inches divided by 12 equals 13⅓ inches.

These sizes are not too different from the actual dimensions or perhaps the smaller beam used on the Lap Engine was in stock!

Apart from any pit sawing or adzing to size, very little extra work was needed to be done to the main beam. All the attachments were either bolted or clamped into their position.

The preventer was also a balance weight on this engine. Engines which used oscillating beams were usually fitted with attachments designed to stop any sudden or uncontrolled movements of the main beam. The first engine to have these devices fitted was Thomas Newcomen's Dudley Castle engine in 1712. These attachments were usually strong timber cross-members which were bolted onto the beam as near as possible to the arch heads. When an engine was running normally, these attachments served no other useful function, than to prevent the main beam from being drawn too far down the well head by the weight of the pump rods.

Also, a situation could arise if a piston rod or its attachment to the main beam failed. These preventers, as they were called in later years, would stop the

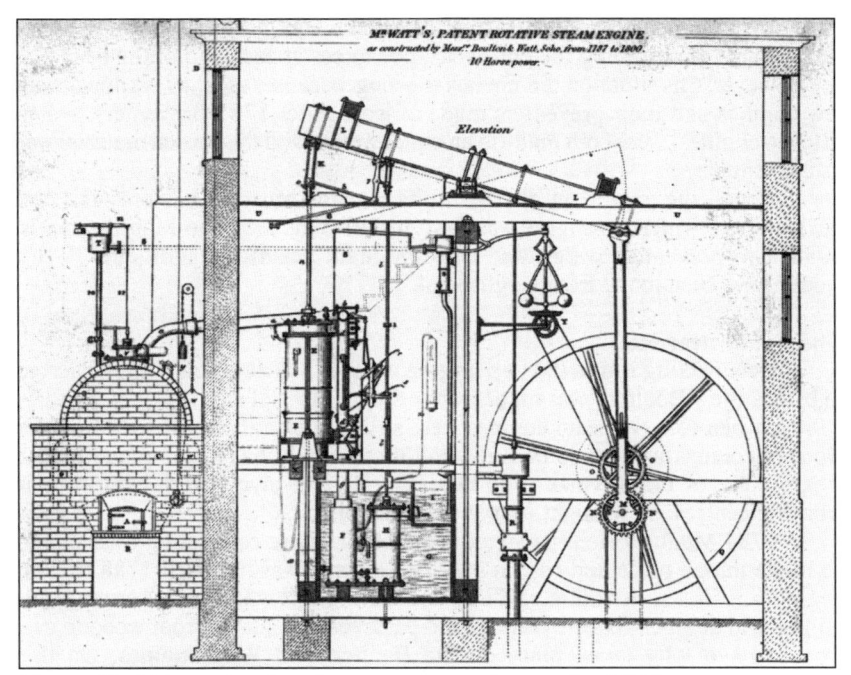

*Figure 63. This engine drawing, represents the standard 10 horse power Boulton & Watt engine which was built between 1788 and 1800 and shows the parallel motion, sun and planet gearing and also the condenser cooling tank. This drawing is from John Farey's book, 'A Treatise on the Steam Engine 1827'.*

rocking beam's descent by making contact with the spring beams and so minimise any further damage.

An even more serious condition could arise if no preventer was fitted for example in the event of a pump rod or arch head chain failure. In such a case the applied load to the piston would suddenly disappear, the piston could then accelerate out of control and break the end out of the cylinder. The attachments which were fitted on to Lap Engine's main beam had all the classic shapes of timber preventers. However, the preventer which was fitted on to the beam at the piston rod end, was made from cast-iron with an approximate weight of 400 lbs.

In the early development of rotative beam engines, it was always considered that the weights on each side of the beam's pivot had to be equal. This preventer with a weight of 400 lbs, then balanced the weight of the connecting rod at the other end of the beam. Perhaps the large-toothed flywheel was thought to contain only enough rotary energy to maintain the engine output shaft at a constant speed of 50 revolutions per minute.

It was not understood at first that the kinetic energy stored in the flywheel was capable of absorbing the out-of-balance forces at each end of the beam.

However, this situation did not last too long because, as far as is known very few engines had their preventers made of iron. After 1788 preventers, when fitted to engines, were both made to an equal weight and the chosen material was usually timber.

Although the engine on display at the Science Museum has only the cast device fitted, when the engine was first installed at the Soho Manufactory in 1788, two preventers were fitted onto the main beam; the additional device would have been made from English oak.

## The Connecting Rod

The connecting rod of the Lap Engine or 'spear' as it is sometimes referred to by Matthew Boulton, was made from a single piece of cast iron. In the early 1780's, when rotative beam engines were still in their infancy, it was usual for wooden connecting rods to be fitted and these would have been made from a single piece of English oak. The two separate end pivots were held at their working centres by wrought iron straps and bolts.

In 1784 Matthew Boulton suggested that all future connecting rods should be made in one piece and cast in iron. However it was not until 1788 that an engine was first seen with a cast iron connecting rod made to his recommendations and another sixteen years had to pass for the change from wood to cast iron to have fully taken place on the Boulton and Watt engines. On this evidence, the fitting of a cast iron connecting rod to this engine appears to be a most advanced step in the design of rotative beam engines.

Perhaps the following theory could help to explain why wooden connecting rods were fitted for such a long time after Matthew Boulton's recommendations. The changeover not really being completed until beam engines had ceased to be an integral part of the building in which they were used. In the early years, rotative steam engines were always an integral part of the building structure and thus both the engine and the engine house had to be erected at the same time. Most of the construction sites would have been away from any manufacturing facilities and at such installations, even when taking the greatest of care, inaccuracies in the final size of the building would occur. This could make the centre distance of the connecting rod pivots difficult to predetermine accurately.

The actual centre distance of the pivot bearings for a connecting rod was relatively unimportant, the only criterion being that its centre measurement must equal the distance between the small end pivot and the centre line of the main driven shaft (or crankshaft) when the main beam was placed in its horizontal position. Obviously this centre distance could not be predetermined accurately, because the measurements in use were only those needed for the construction of the engine house.

Both the small and big end components were supplied by Boulton and Watt, so it was a simple operation to place the small end and big end bearings in their working position and to cut the wooden spacer to suit this distance after the

## STRAIGHT LINE MOVEMENT FOR A PISTON ROD

building was completed.
 Some long delays would have taken place if this method had not been used. The alternative was to make a pattern and cast the rod when the engine installation was almost finished, but it was possible that at these engine sites there were no casting facilities available. However the Lap Engine was erected on a site where there were extensive manufacturing facilities and so there would be few delays in the making of the cast iron rod to complete the engine's installation.

A portable rotative beam engine was the first engine to use a connecting rod made from cast iron in May 1788. This was a small engine with a double action cylinder 10 inches in diameter. Many engines built after c1800 were constructed as a single unit and did not depend on the actual engine house for support.

Perhaps this could help to clarify the theory because, at this date, an almost complete change over to cast iron connecting rods had taken place. The working centres of the connecting rod using this method of construction could be predetermined before any assembly had taken place.

**The Upper End of the Connecting Rod and how it was Secured to the Main Beam**

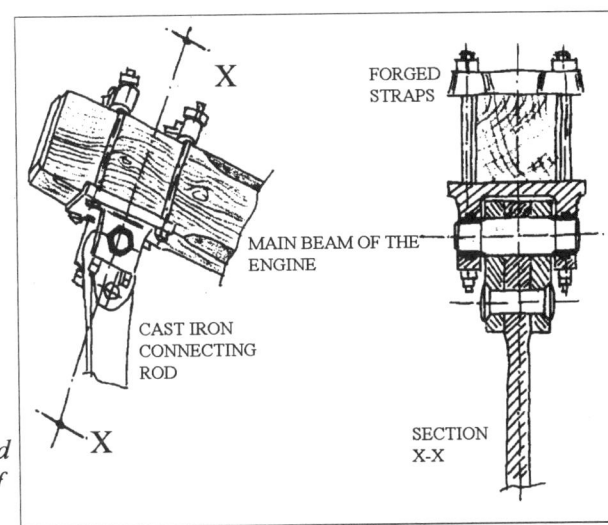

*Figure 64. How the connecting rod, or 'spear', was secured to the main beam of the engine.*

At the upper end of the connecting rod used on the Lap Engine, there was positioned a small end pivot which is shown in Figure 64. This shaft was secured into the connecting rod by a 'shrink fit' of the two wrought iron side plates, which were first heated, then positioned over the central portion of the shaft. On cooling, the side plates would contract and firmly hold the small end shaft into its working position. The two wrought iron plates and the connecting rod were then secured together by means of a large rivet passing through the whole assembly, giving the small end shaft great strength. Each end of the shaft pivoted in split bronze bearings, held into their position by small nibs cast on the outer face of the bearing shells. The bearings were prevented from turning in the cast housings by the outer edge of the bearing shells being shaped in the form of a hexagon. When the whole assembly of the small end bearing was completed, it was cramped to the main beam using ¾ inch diameter, square headed nuts and bolts.

The connecting rod of the miniature Lap Engine was made after assembling all the component parts together and the final working centres were determined by measuring the centre distance on the miniature in exactly the way as described for the full sized engine.

# CHAPTER 8
# ROTARY MOTION - THE SUN AND PLANET GEARS - GOVERNOR

**James Watt's rotary motion without a crank mechanism.**

Rotary motion without the crank appears to have stimulated an intellectual challenge in Watt's mind and he quickly set himself the task of obtaining rotary motion without any infringement of what he called that 'hateful Pickard patent'. On the 23rd February 1782, a patent was granted to James Watt which covered five different methods, of converting the reciprocating motion of a steam engine's main beam into the rotary motion of an output shaft. Only his fifth method is to be described in detail, as his four other methods of obtaining rotary motion have been given with the details of the Pickard and Wasborough engine.

After all his experiments, the ideal solution was to be found by using what became known as 'sun and planet' gearing. These two gears efficiently converted the reciprocating motion of a steam engine into a rotary motion of the driven shaft. This was the first recorded use of what we now call epicyclic gearing and was used on the Lap Engine to provide rotary motion without any infringement of James Pickard's patent. In the following letter, which was taken from the Boulton and Watt papers, James Watt explains to Matthew Boulton the principle of his sun and planet gears.

'I wrote to you on the 31st, since then I have tried a model of one of my old plans of rotative engines'.

It has the singular property of going twice round, for each stroke of the engine and may be made to go oftener round if required without additional machinery'.

It is popularly believed that one of Boulton and Watt's employees, William Murdock (1754-1839), was the originator of the sun and planet gear, but to this day it has never been clearly established who was the actual inventor. However, with the fitting of this rotary element, the Watt engines could now make rapid progress, and with the introduction of the double-acting cylinder, power could now be transmitted to the main beam on both the upwards and downwards strokes of the piston, which could also power the connecting rod in both directions - up and down.

The double acting cylinder therefore represented an important step forward compared to Newcomen's single action principle because, when a Newcomen engine was producing rotary motion, only 180 degrees of a revolution was driven, it was the energy which was stored in the flywheel which was used to complete a revolution.

**The Operating Principle of the Sun and Planet Gears (Epicyclic Gears)**

Between the sun and planet gear is a connecting link which holds the two gears at a fixed centre distance. Because of the oscillating movement of the main beam, the connecting rod of the engine, with the planet gear rigidly attached to it, is compelled to revolve around the axis of the sun gear.

# ROTARY MOTION - THE SUN AND PLANET GEARS - GOVERNOR

*Figure 65. The sun and planet gears of the Lap Engine in the Science Museum, London. The wooden rule shown inverted was used as a reference scale when sizing components for the miniature engine.*

    With the sun gear rigidly attached to the engine's driven shaft (output shaft), any movement produced by the planet in its orbit around the sun gear is true rotary motion. Rotation produced by this means not only avoided any patent litigation but it also had an added bonus that as the sun and planet gears which were used had an equal number of teeth, this multiplied the output speed of the driven shaft by two. Thus for every oscillating cycle of the engine, the output shaft revolved twice. This occurred because the circular path traced out by the planet gear centre was twice the distance traced out by the circular pitch of the sun gear. Since both the sun gear and the planet gear *must* travel the same circumferential distance, this compelled the driven output shaft to revolve at 50 revolutions per minute, and not at the 25 oscillating cycles per minute of the engine.
    If a crank had been used to produce rotary motion on the engine the main shaft would have only revolved at 25 revolutions per minute. An additional set of gears would have been needed to produce a suitable working speed for the lapping and polishing machines at the Soho Manufactory.

# ROTARY MOTION - THE SUN AND PLANET GEARS -GOVERNOR

*Figure 66. The sun and planet gears as made for the miniature engine showing the connecting link.*

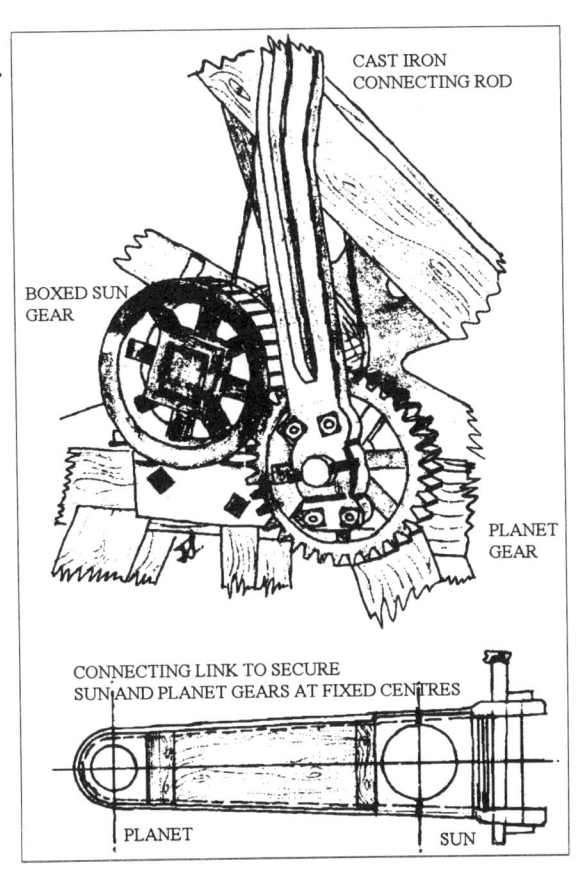

## Constructional Details of the Sun and the Planet Gears

Both of these gears were made of cast iron and each had 32 teeth, with a face width of 2½ inches. Both gears were used 'as cast' from the foundry, though the bore of the planet gear could have been machined to size on a lathe.

The planet gear was held in position onto the lower end of the connecting rod by a sturdy axle, and was retained by a square dowel which was driven through the end of the axle. Two strong 'U' bolts were also used to clamp the planet gear securely into its working position and these bolts also prevented any rotation of the gear around the end of the connecting rod.

The sun gear was secured to the engine's output shaft by sixteen tapered wrought iron wedges. This method was used to avoid any need for accurate machining on the gear. Hopefully, once the gear was set in place, it never slipped. A square bore was cast centrally into the sun gear, having a clearance of ½ inch over the size of the square main shaft. This clearance was then used

# ROTARY MOTION - THE SUN AND PLANET GEARS -GOVERNOR

to provide an adjustment for the true running of the gear by the use of the tapered wedges, which were simply driven in or pulled out until the gear was running as true as possible.

The sun gear was an 'encased' or 'boxed' gear, the essential reason was to guide the teeth of the planet gear in their orbital path around the sun gear and, to prevent any lateral movement. This also added greater strength to the teeth of the gear, possibly to prevent any breakage. In order to hold both the sun and the planet gear at a fixed working centre a connecting link was used. This simple construction consisted of two split-bronze bushes held apart by a wooden spacer, all of which was surrounded by a wrought-iron strap held in position by a traditional gib and cotter.

The connecting link was not attached structurally to the engine in any way. Arranged in this way James Watt achieved true rotary motion without any infringement of James Pickard's patent and, the Sun and Planet method was used on all Boulton and Watt engines until 1792, when Pickard's patent finally expired.

## The Flywheel

When James Pickard's Newcomen type pumping engine was adapted to provide rotary motion in 1780 the flywheel was used to store enough kinetic energy to maintain the engine at a constant rotational speed. James Watt did not appreciate the full effect the flywheel contributed to the rotary motion provided by a steam engine, but he did give credit to its originator, Matthew Wasborough who was the man responsible for the fitting of the flywheel to James Pickard's engine.

On James Watt's newly developed double acting engines the flywheels were only an initial refinement however they greatly added to the uniform nature of the rotary movement. The flywheels were used as a reservoir of energy which overcame any fluctuations in the workload applied to the engine and were not now required to keep the engine in continuous motion.

## The Toothed Flywheel which was fitted to the Lap Engine.

The toothed flywheel on this 10 normal horse power engine was the largest wheel ever to be fitted to an engine of such low power. The large diameter was designed to accommodate the great number of teeth needed to produce a high gear ratio between this geared wheel and the driven countershaft, which was then used to drive the lapping and polishing machinery at the Soho Manufactory. Lapping and polishing machines need to be driven at a high speed, but most steam engines of this date (1788) usually drove low speed machinery such as grinding pans, corn mills and rolling mills. Exactly how the engine transmitted its power to the countershaft is not known. However, the arrangement reproduced on the miniature engine seems practical and was probably the method used.

The outside rim of the flywheel contained 304 wooden teeth equally spaced on a pitch circle diameter of 15 feet 8 inches, these teeth were made of hornbeam.

# ROTARY MOTION - THE SUN AND PLANET GEARS -GOVERNOR

*Figure 67. The sectional flywheel on the engine in miniature. Showing the wooden gear teeth.*

304 WOODEN TEETH ON 15FT 8 INS PCD

ALL CAST IRON

OPEN TOP TO BEARING

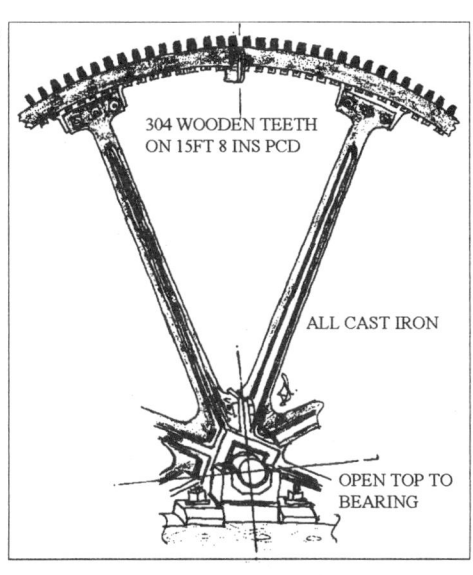

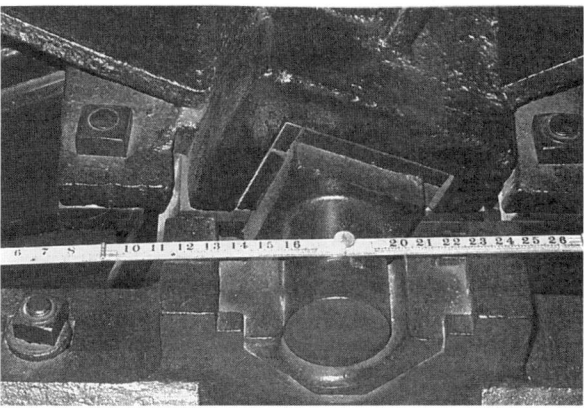

*Figure 68. The open top to the bearing. The great weight of the flywheel made it unnecessary to fit a cap to the bearing. The ruler is for scaling purposes.*

    This timber was chosen because of its high strength and its resistance to any splitting. The wooden teeth were held in their cast sockets by a ¼ inch diameter wrought-iron dowel. The outside rim, or annulus, was constructed of eight cast-iron segments held together by sixteen ⅝ inch diameter square headed nuts and bolts. The central hub had a 7 inches square cast bore and the whole assembly was secured onto the engine's 5½ inches square driven shaft by sixteen wrought iron wedges.
    This complete assembly with the sun gear rotated in large plummer blocks bearings, all supported by plain, bronze bushes. The bearing immediately behind the sun gear was of a conventional construction, using two split-bronze shells

# ROTARY MOTION - THE SUN AND PLANET GEARS - GOVERNOR

with a diameter of 5 inches. However at the other end of the main shaft only the lower half of a bearing was used. It must have been considered that the great weight of the flywheel, in the region of 4,500lbs was more than adequate to hold this shaft in its place, without the addition of a bolted on cap.

## Rotary Speed Regulation

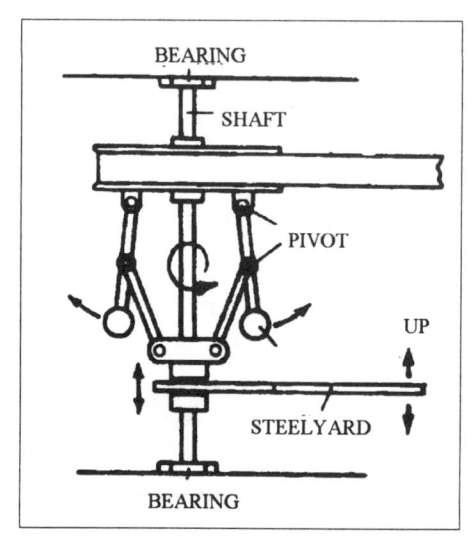

*Figure 69. The governor used at the Albion Mill in London which controlled the force between the mill stones. This device was patented by Thomas Mead in 1787.*

It was not until 1788 was drawing to a close that the rotary speed of a steam engine could be controlled by automatic means. Before this date the rotational speed of any engine had to be carefully judged by the engineman, whose responsibility it was to open or close the valve which regulated the volume of steam entering the condenser. This manually operated valve was fitted into the main eduction pipe from the boiler and into the condenser then the valve was opened to increase the speed and closed to reduce the speed of the engine. This would have been a very demanding job because when the engine was running it could not be left unattended, either through the day or night.

Matthew Boulton in 1788 visited the Albion Mill in London, specifically to look at an earlier Boulton and Watt engine, this was a large engine which was powered by a double acting cylinder. To Boulton's surprise in another part of the mill a device had been fitted which controlled the milling pressure between the upper and lower stones on a corn grinding mill. This device was made up of two lead weights attached to the end of light rods which were then attached to a central rotating shaft. The rods were allowed to pivot freely at their anchored end. The central shaft was rotated by the means of a flat leather belt which was driven by the main rotational part of the mill.

# ROTARY MOTION - THE SUN AND PLANET GEARS - GOVERNOR

*Figure 70. A copy of the original drawing showing the governor which was fitted to the Lap Engine. The drawing is dated 13th December 1788.*

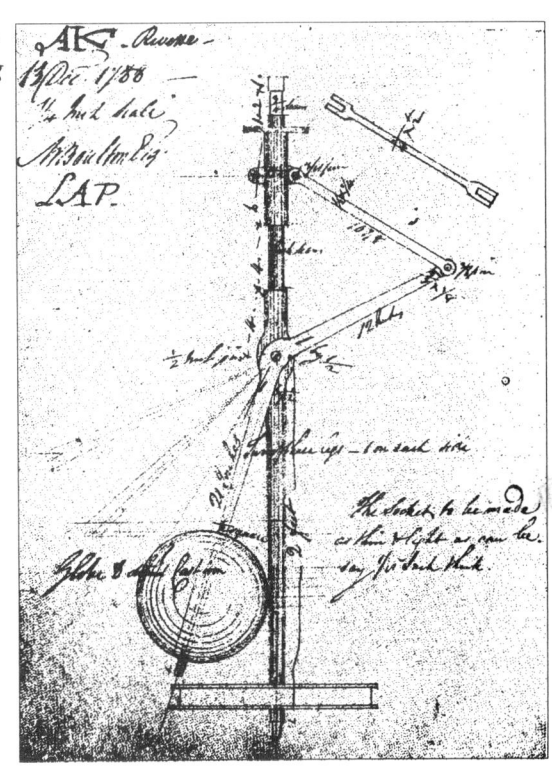

The centrifugal force which was generated by the varying rotational speed raised or lowered the central collar of the whole device. The collar was attached to a steelyard which operated the tentering screw, which increased or decreased the force between the corn grinding stones. Matthew Boulton quickly appreciated that this governing device had great potential for controlling the rotational speed of a steam engine and he wrote from London telling James Watt of his discovery.

The milling machinery at the Albion Mill had been installed by a leading Scottish designer, Andrew Meikle (1719-1811), and it was Meikle who fitted the centrifugal devices to control the operation of the milling stones.

He possibly had this control idea from a patent which was granted to Thomas Mead in 1787. This patent reads:

'Thomas Mead of Port Sandwich, Kent Carpenter. A regulator on a new principle for wind and other mills, for the better and more regular furling and unfurling, the sails on windmills without the constant attendance of a man, and for grinding corn and other grain, and dressing of flour and meal superior in quality to the present practice, and for regulating all kinds of machinery where the first power is unequal'.

105

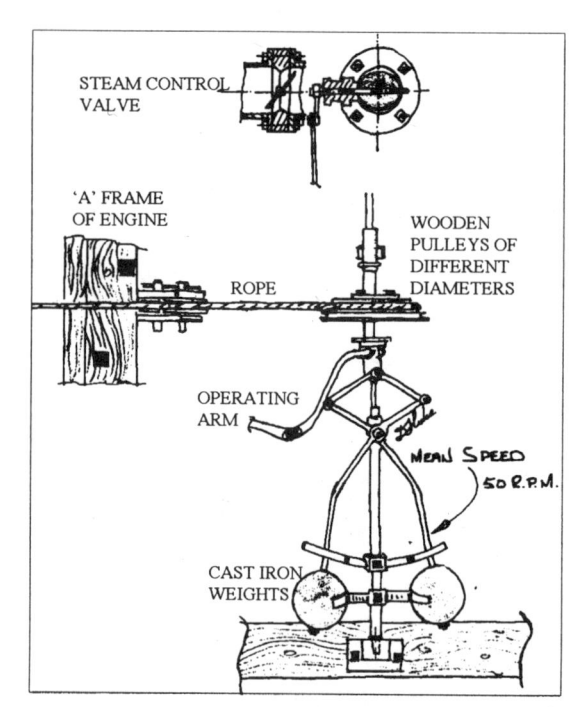

# ROTARY MOTION - THE SUN AND PLANET GEARS - GOVERNOR

*Figure 71. The governor which was fitted to the model of the engine.*

When Matthew Boulton returned from his visit to the Albion Mill, development work at the Soho Manufactory went ahead very quickly as Boulton and Watt wanted to control the speed of their rotative beam engines with this new centrifugal device. By 13th December 1788, a drawing was completed showing a centrifugal governor destined for use on the Lap Engine. This drawing Figure 70 clearly shows the operating principle, which was derived from the mill governor, as both governors were driven by the means of flat leather belts.

When this automatic speed control was fitted to the Lap Engine, it was seen as the world's first rotative steam engine ever to use an unmanned speed control.

Speculation can only decide the first date at which the Lap Engine had its speed controlled by this new automatic means.

The drawing of 13th December 1788, would not have allowed enough time to construct and fit the governor before the beginning of the year 1789, or could the drawing have been produced *after* the actual governor had been made and tried out on the engine? This practice was widely used in the eighteenth century because it allowed new ideas to be perfected on the actual experimental machinery before being drawn out properly. This drawing could have been the final recording for future reference to make governors for subsequent engines, but this can only be supposition!

# ROTARY MOTION - THE SUN AND PLANET GEARS -GOVERNOR

## The Function of a Governor

A governor is required to control the mean speed of the engine over a period of time, as distinct from the flywheel which minimised the fluctuations of speed during one revolution of the engine.

A decrease in a load on the engine was accompanied by an increase in speed and the governor was then required to operate so as to decrease the supply of steam to the condenser, thus bringing the speed back to its original value. The converse action that was required following an increase in load was achieved by connecting the rotating parts of the governor through suitable levers to the controlling lever of the throttle valve.

## The Governor of the Lap Engine

At some time during the engine's long working life, the original device was replaced by the governor shown in Figure 71, but the date when this replacement was fitted can again only be speculation. However, a device of almost identical proportions to the replacement governor was fitted to an engine designed by James Watt in 1798 and this was driven from the engine's rotary output shaft by an endless rope.

Where the original governor was positioned on the engine in 1788 is not known, but great difficulties would have arisen if the position was the same as its replacement because the flat belt needed for driving it would have been very long and difficult to guide around the outside edge of the flywheel.

The rope which was needed to drive the governor wound its way through an awkward passage to the wooden pulley of the governor which had two grooves on different diameters. This allowed the rotational speed of the governor to be adjusted for the average speed to be set. A series of wooden pulleys guided the driving rope around the outside edge of the flywheel.

These early 'Watt governors', as they were to become universally known, were very sensitive to any fluctuation in speed and when fitted to this engine, fully opened or closed the throttle valve with a variation in speed of only twenty-five revolutions per minute.

# ROTARY MOTION - THE SUN AND PLANET GEARS -GOVERNOR

*Figure 72. A photograph of the engine in miniature showing the flywheel, governor and main powering cylinder, all to the scale of one sixteenth full size.*

# `CHAPTER 9

# HOW THE ENGINE WAS ASSEMBLED

### The Wooden Frame of the Lap Engine

The complete structural framework of the engine was assembled from wood all held together with wrought iron straps and bolts. James Watt stated that the best English oak should always be used as this was easy to obtain and it also possessed great strength. An interesting example of the methods used for the construction of the wooden framework is shown in Figure 73.

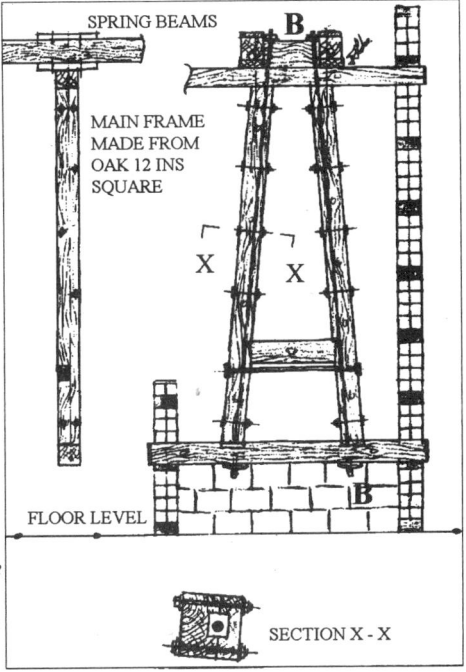

*Figure 73. The wooden 'A' frame support, showing how it was held together with long bolts. Shown as B-B.*

The main 'A' frame was held together between the upper and the lower support beams by two 18 feet long wrought-iron bolts of 1 inch in diameter. Each upright beam of the 'A' frame was constructed in two separate parts (see section XX) with a groove cut down the full length of the adjacent faces to allow the long bolts to clamp the frame together. The largest baulks of timber which were used in the framework of this engine were adzed to their correct size and the smaller pieces were cut to the correct size by pit sawing. Great care was required in the selection of this timber and only well-seasoned heartwood was recommended. Oak which has not been well seasoned possesses a very corrosive sap (tannic acid) which would have rusted the iron parts of the engine.

So that a stable structure could be built from these (relatively) flexible frames,

## HOW THE ENGINE WAS ASSEMBLED AND OPERATED

the long horizontal wooden beams at the top and bottom of the frame were built into the end and side walls of the engine-house.

### Controlling the Valve Gear

The general arrangement drawing Figure 74 shows how many of the main cylinder's operational components were fitted together, from the cylinder head, to the base of the separate condenser positioned inside the wooden tank. All of the component parts shown on this drawing have been described in preceding chapters and they can be seen assembled together in this pictorial drawing. Hopefully the control and the function of the main powering cylinder to this engine can now be easily understood.

From the boiler, the flow of steam into the condenser was controlled by a throttle valve which was opened or closed by the levers connected to Watt's centrifugal governor. By studying this diagram, it will become clearer how these levers communicated any movement from the sliding collar of the governor to the throttle valve. It was this valve which controlled the flow of steam first to the cylinder head and, then into the condenser positioned below the cylinder.

The elliptical plate (d) still causes confusion to many observers - why is it made in such an odd shape? Research into this engine has not revealed why this plate was fitted above the sliding collar of the governor. The drawing also shows the control of the valves inside the nozzle housings. The volume of steam passing through these valves and into the cylinder was controlled by leather straps and buckles which are shown at (a) on this drawing.

The vacuum gauge (e), positioned between the governor and the main cylinder, was connected by a small copper tube which passed through the lid and into the condenser. This simple device consisted of a free moving piston which slides inside a brass cylinder, the piston being drawn by the vacuum and pulling against a spring-loaded pointer. The pointer showed the degree of vacuum by its movement against a calibrated brass quadrant. The reading was graduated in inches of mercury.

Positioned close to this vacuum gauge was a brass stop and start lever, which was connected with a light iron rod to the water injection valve. The injection valve was positioned on the outside of the condenser wall and was used to regulate the water spray which was used to condense the steam and create the vacuum.

A little in front of the condenser is the air pump which removed the accumulation of water from inside the condenser. This was completed on every powering stroke of the engine.

Seen also inside the large wooden cooling tank are the valve operating weights. In the description of the valve chest and nozzle housings, it was described how these weights were submerged in the cooling water to moderate the oscillating and swinging movements when the engine was working at full speed.

The whole timber structure supporting the cylinder and the condensing tank was also built into the walls of the engine house.

## HOW THE ENGINE WAS ASSEMBLED AND OPERATED

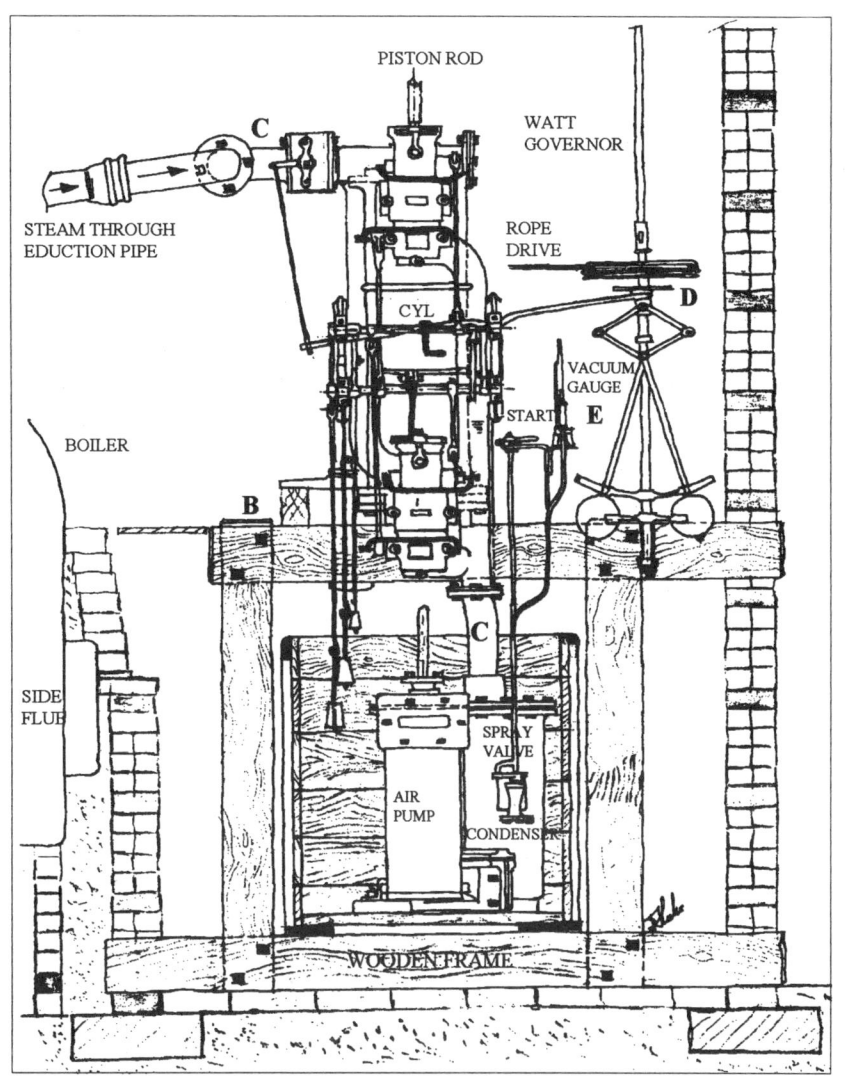

*Figure 74. A sectional view of the Lap Engine showing how many of the component parts are fitted together. For additional support, the beams are built into side walls of the engine house.*

## HOW THE ENGINE WAS ASSEMBLED AND OPERATED

The main cylinder was held into its working position on the wooden frame by two large bolts and wrought iron straps.

The nozzle housings can be seen at the top and bottom of the cylinder together with the ingenious 'G' clamps retaining the valve operating spindles (c) shown on Figure 58.

## Some General Information for the Efficient Running of a Rotative Steam Engine supplied to each customer who purchased an engine of 10 HP

### General Lubrication

Melted tallow was used for the lubrication of the piston and cylinder and was also recommended for the piston rod seal (stuffing box).

Spanish olive oil was used as a general all-purpose lubricant, and was sometimes thickened by the addition of melted tallow or butter. This mixture was recommended for all the small pivots, and was also recommended for the rotational shafts supported in the plummer block bearings.

John Farey in his *'Treatise on the Steam Engine'*, stated that linseed oil should never be used as this oil dries too hard and increases the friction which, in its turn, reduces the power output from the engine. Also the bearings had to be cleaned well before any fresh lubrication was carried out.

The weekly allowance of tallow for lubrication of the cylinder and piston was stated as 3lbs for an engine of 10 horsepower.

### The General Maintenance of the Stuffing Boxes and the Piston

Some extracts from Boulton and Watt's operating instructions are reproduced below.

1. ' A very important article is the proper packing of the piston. Take sixty white or untarred rope yarns, and with them plait a gasket or flat rope, as close and firm as possible, tapering for eighteen inches at each end, and long enough to go round the piston and overlap for that length; coil this rope the thin way as hard as you can, lay it on an iron plate and beat it with a sledge hammer until its breadth answers its place; put it in and beat it down with a wooden driver and a hand mallet; pour some melted tallow all round then pack in a layer of white oakum, about half an inch thick, then another rope and more oakum, so that the whole packing may have the depth of four inches, or only three inches if the engines are a small one; soak the whole well with melted tallow, and after having beat the packing moderately, lay on the piston head; put on the springs and screw them down. In a new engine the piston must be examined after about twelve hours going, and be beat a little and fresh greased; but you must be careful not to pack or beat it too hard; otherwise it will create so much friction and almost to stop the engine.

2. ' The buckets of the air and hot water pumps are to be packed with a flat rope, wrapt round them, edge ways. The ends of these gaskets should be made fast, by being drawn through holes made in the buckets for that

## HOW THE ENGINE WAS ASSEMBLED AND OPERATED

purpose, and secured there by wooden pegs hard drove in. The gaskets should be well smeared with tallow before the buckets are put in, and they should not fit the pumps too tight, as their sticking is very troublesome especially at first.

'The stuffing boxes of the cylinder and air pump are to be packed by wrapping a soft rope round the rod. And beating it in until it nearly fills the stuffing box remembering to soak it well with tallow as you go on; above this rope lay on the wooden collar and screw the gland down upon it moderately tight.

3. 'Once every week let the top of the cylinder be taken off, and also the springs and head of the piston, let the packing be beat down moderately, with the driver and mallet, and fresh oakum, or a gasket added when necessary. For every foot the cylinder is in diameter pour two pounds of melted tallow on the packing'.

### 4. The Recommended Jointing Material.

'To make putty for making or repairing the joints, take whiting, or chalk finely powdered, dry it on an iron plate, or in a ladle, until all the moisture is exhaled; then mix it with raw linseed oil, and beat or grind it well, adding more oil or whiting, until it is of the consistence of thick paint and perfectly free from lumps or inequalities'.

'For some purposes, where the putty is wanted today and to be very sticky, use painters drying oil, which is made by boiling the oil with a small quantity of litharge or red lead'.

5. **The General Condition.** 'When an engine is in tolerable good order it will bear to stand ten minutes, and go to work again without blowing afresh, and though it has stood two or three hours, if there has been any steam issuing from the boiler, and no air has been admitted into the cylinder, it will generally go off with once blowing for about a minute'.

6. **Correct Boiler Care.** 'Let all the coals employed to feed the fire, be thoroughly watered just before they are thrown on as that will prevent their being swept into the flues by the draught of the chimney'.

'The fire should be kept of an equal thickness and free from open places or holes, which are extremely prejudicial, and should be filled up as soon as they appear; if the fire grows foul and wants air by clinkers collecting on the bars, they must be got out with a poker, but the fire should be as little disturbed in that operation as possible, and the greatest care taken not to make any coals or coaks falls through, which are not thoroughly consumed; it is very common for a fourth of the whole coals to be wasted in this manner, by mere carelessness,

'When the fire is newly made, the damper should be raised a little, so as to let off the smoke freely, but should be let down to its proper place so soon as the smoke is gone off. The air door in the chimney should be always open more or less; it prevents the flame from being sucked up the chimney, and very considerably increases the effect of the coals. Once a month, the boiler and flues ought to be cleaned, or oftener if the waters are very subject to incrust the boiler'.

'Every morning the ashes ought to be taken out of the engine house swept clean, and a view taken of every part of the engine, to see that nothing is working out of its place, or want oiling'.

## How the Lap Engine Operated

The water was heated in the waggon boiler and was allowed to build up to a working pressure of 5 inches of mercury, ie. about 2½ pounds per square inch. From the boiler, the steam was transferred along the 5 inches diameter eduction pipe to the operating cylinder, and from the cylinder into the condenser. On entering the condenser the steam had further cooling in the form of a water spray created by the inherent pressure of water in the wooden tank. This spray was controlled by an injection cock or valve. The spray of water condensed the steam and a vacuum was created.

The vacuum was then directed to the cylinder by means of the nozzle housings positioned at the top and bottom of the cylinder and so the vacuum powered the piston in both directions to produce linear motion of the piston rod.

The nozzle housings fitted to this engine contained two drop valves in each, enabling the admission of low pressure steam to be applied to one side of the piston and a vacuum to the other. The low pressure steam was used to maintain the powering cylinder at a constant working temperature.

The vacuum created was always chilled by the cold water surrounding the condenser, while the steam on the 'pressure' side of the piston remained hot. If the working cylinder was not maintained at a high temperature and was powered only by a vacuum, an eventual hydraulic lock would occur caused by an accumulation of water inside the cylinder.

The vacuum was delivered alternately to each side of the piston by a complicated system of levers attached to the nozzle housings, or chambers. These levers opened and closed the drop valves which were held firmly to their brass seats by weights attached to the end of light wrought iron rods, with the valves being operated by two adjustable wooden tappets secured to the wooden plug tree which moved up and down. The tappets worked at two levels, one on the upward stroke and one on the downward stroke.

After each cycle of the engine, the water from the condenser was removed by the air pump which then flowed into a cast-iron receiving tank fixed inside the main cooling tank.

As air could enter the condenser with the steam from the boiler, a brass snifting valve was fitted near the bottom of the condenser. This valve allowed the excess air and steam to escape to the atmosphere, but still contain the vacuum which had been created after condensation by the water spray. The water then flowed from the condenser through a one way valve into the air pump.

The water was then lifted by the air pump into the cast iron receiving tank, and any excess water not required by the engine then passed over a weir which maintained a fixed level within the tank. The bulk of the water in the cast iron tank was lifted by means of the secondary pump to a header tank attached to the

## HOW THE ENGINE WAS ASSEMBLED AND OPERATED

main frame of the engine, from where it dropped by gravity to the header tank of the boiler. The water from this header tank then refilled the boiler.

It was found that more water was always pumped back than was lost by means of evaporation. This was because of the additional water that had been added to it from the condensing water spray. The surplus of water was then returned to the main wooden condensing tank. A considerable amount of energy was saved by recirculating the water from the condenser as the water was still at a temperature of about 85 degrees fahrenheit and so the boiler was not chilled when it was refilled.

During the operation of the engine the condenser tank had to be maintained at a constant level and a low temperature and this was achieved by the primary pump drawing its water from Hockley Brook which flowed from Soho Pool at the Manufactory. Water was always drawn in excess of the required amount by the primary pump, so the surplus then overflowed back into the brook. This action of flowing water greatly increased the efficiency of the engine because it maintained the condenser tank at a consistently low temperature which was essential for the good condensation of the steam.

# HOW THE ENGINE WAS ASSEMBLED AND OPERATED

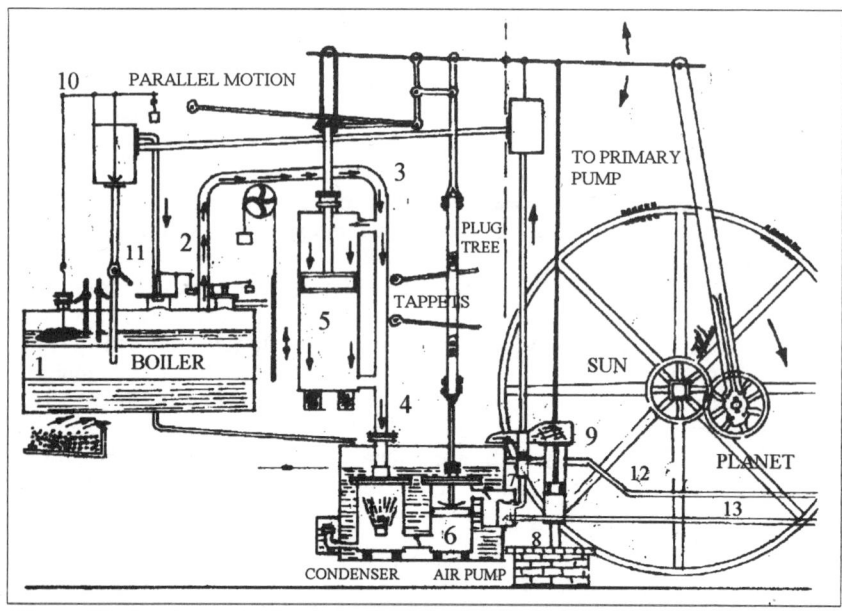

*Figure 75 A schematic drawing showing the flow of steam from the waggon boiler and into the condenser.*

1. The steam is first of all raised to 2½ pounds per square inch in the waggon boiler.
2. The steam is then transferred through the 5 inch diameter eduction pipe to the powering cylinder.
3. The steam reaches the top of the cylinder
4. The steam is then transferred into the separate water cooled condenser.
5. A vacuum is created inside the condenser by a cold water spray.
6. An accumulation of condensate and water spray is removed from the condenser by means of the air pump.
7. The water from the air pump then flows into the cast iron receiving tank.
8. Water from the receiving tank then flows over a weir and then to waste.
9. Water from the receiving tank is then lifted by the secondary pump into the header tank of the engine, ready to refill the boiler.
10. From the engine's header tank the water now flows into the boilers header tank by the means of gravity.
11. The boiler is refilled by a standpipe which is controlled by a flow regulator
12. The primary pump fills the condensing tank with cold water from Hockley Brook
13. Excess water from the condensing tank which has flowed over the weir then flows away to waste.

# CHAPTER 10
# OPERATING INSTRUCTIONS

**Starting of the Lap Engine into Rotary Motion from a Stationary Position**

A strict procedure had to be followed when the engine was started from rest with an empty boiler. Fresh water from the Hockley Brook was first of all pumped into the waggon boiler. Portable hand pumps were most probably used, although no reference as to the type which were employed could be found. Water entered the boiler from the header tank after the balance weight had been removed. The water then flowed down the delivery pipe after fully opening the flow regulator. The working capacity of the boiler was approximately 360 gallons and, as the boiler filled, the balance weight was periodically replaced to see when the correct water level had been achieved. With the water level correct, the fire could now be kindled. The flue damper was set at two-thirds closed and the small inspection door at the base of the chimney opened.

Only with a combination of these two factors could the draught on the fire be regulated enough for the fire to be lit without drawing the kindling fuel into the side flues surrounding the boiler. Three hours were usually needed to bring the water to the boiling point, and during this time, the engine man had removed the weight from the pressure safety valve which ensured that the air space over the water was completely replaced by steam. Also during this time, the engine had to be manoeuvred into its ideal starting position by barring over the flywheel until the planet gear was set at the 3 o'clock position.

The safety valve weight was then replaced, enabling the steam pressure to be raised. With the fire burning well, the damper could be fully opened, and the small door at the base of the chimney closed. In order to achieve the working pressure of 5 inches of mercury on the steam gauge, ie. 2½ pounds per square inch it generally took another fifteen minutes after the closing of the safety valve. The change over levers which controlled the steam valves, could now be operated by hand. These levers were centrally positioned, thus allowing all four drop valves to be open.

With all the valves open and the injection valve closed, steam was allowed into the cylinder and through into the condenser. This was known as, - to quote from John Farey's book, *A 'Treatise on the Steam Engine 1827'*, 'blowing through,' and had to be done to remove any build up of water from the eduction system through to the air pump.

Once this had been completed, the water spray was introduced by opening the injection valve and allowing the water spray to condense the steam and form a vacuum which could then be used to power the engine. The degree of vacuum was checked with a fixed manometer situated by the cylinder head. When the reading was approximately 20 inches of mercury, the tappets could be returned to the automatic running position. With the help of two or more strong men, the flywheel could now be set into motion.

If the correct starting procedure had been followed, the engine would

## OPERATING INSTRUCTIONS AND POWER CALCULATION

continue to revolve and it would then be left under light loading until the powering cylinder had reached its ideal working temperature.

It was usual for the engine to start at once, but if the start had not been successful, the 'blowing through' procedure was repeated with the manual valve setting once again. With the engine in motion, a constant speed of 25 cycles per minute was maintained by Watt's centrifugal governor. The engine operated at 25 strokes per minute rotating the driven shaft at 50 revolutions per minute, because, of course, the final shaft was driven by the sun and planet gears.

### Some Thoughts on Sealing the Steam Pipes

Dissatisfied with the previous methods which were used to join steam pipes together, James Watt sought a quicker and easier way. Most joints had been made by clamping lead rings covered with glazier's putty between steam pipes with flanged ends. This method had been in constant use for about sixty years and was also used for sealing the joints on Newcomen type engines. In 1782 James Watt was inspired by seeing the iron foundry practice of filling flaws in castings with a mixture of fine sand and iron filings, moistened with urine! After much thought and experimentation, this practice was used to seal all the steam joints on the Boulton and Watt engines and, for the first time, effective and quick male and female joints were possible. This newly developed sealing substance was called 'iron cement'.

Watt had tried for two years to develop this cement without much success. In 1784, on hearing that William Murdock had successfully developed the substance independently in Cornwall and was using it with great success on all his installations, James Watt ceased his experimentation and adopted Murdock's formula on all subsequent Boulton and Watt engines. Watt's 'iron cement' was made from the metal cuttings (swarf) left after boring cylinders, and the swarf was mixed with sal ammoniac or urine which promoted rapid rusting. The expansion in the joint caused by the rust enabled steam tight joints to be made. When engines were being delivered to a customer's site, the cylinder borings from John Wilkinson's Bersham Ironworks were also despatched. This allowed the 'iron cement' to be made when the engines were being erected.

# OPERATING INSTRUCTIONS AND POWER CALCULATION

## How the Power Developed by a Rotative Steam Engine was Measured

In preceding chapters, the term 'horsepower' has often been mentioned, this term equating the power of an engine to a horse has been in constant use from the very early days of steam power right up to the present day. It is only with the widespread use of standard metric terminology that 'horsepower' is gradually being replaced by 'kilowatt units'.

When steam engines were in their infancy, the power they produced was of an unknown quantity as they were only used to perform a specific task such as the pumping of water. This was quite suitable when the engine was performing a task which had never been accomplished by any other method. However, with the advent of the double acting rotative steam engine, work could now be undertaken which had previously been carried out by harnessing the power of the horse.

Customers naturally wanted to know how many horses the steam engines could replace. James Watt gave this question much thought and decided to measure the amount of work which the average horse could carry out in a one minute period. One day a prospective customer from Manchester informed Watt that the average mill horse harnessed to a shaft could walk in a circle of 24 feet in diameter and could rotate the shaft two and a half times in one minute.

While the horse was doing this, it was capable of holding a rope with a tension on it of 172 lbs. Upon multiplying the distance the horse had to walk in a minute by this average pull, the universally accepted term 'foot pounds' (lbs/ft) evolved. This calculation, worked out at 32,400 lbs/ft but was later raised to 33,000.

*Figure 76. The Soho Foundry was built in 1796 to manufacture the complete steam engine and continued to make engines until 1895.*

# OPERATING INSTRUCTIONS AND POWER CALCULATION

Thus it can be acknowledged that the work carried out by the horse is equivalent to the raising of a weight of 33,000 lbs through the height of one foot in one minute. Therefore, the measurement of the horsepower of an engine is the work done in one minute by the moving piston of an engine, and all divided by the constant number of 33,000.

## The Soho Foundry and the Soho Manufactory

The Soho Foundry was opened on the 30th January 1796. Confusion sometimes arises between the Soho Foundry and the Soho Manufactory, but they were two separate establishments existing approximately one mile apart.

The Soho Foundry was the first factory ever to be built for the sole purpose of producing a complete steam engine and it continued to produce Boulton and Watt steam engines until 1895.

The Soho Manufactory never made the *complete* steam engine, however some of the smaller component parts, such as the lathe turned valves, were produced here, while the valve operating levers were forged in the manufactory's smithy. The wooden patterns used to cast the nozzle housings were also made at the manufactory by the Boulton and Watt pattern makers. However, the actual castings were produced by other specialists - for example John Wilkinson's Bradley Ironworks cast the nozzle housings.

The Soho Foundry is still in existence and is now the factory of W & T Avery Ltd (scale manufacturers), but the Soho Manufactory ceased working in 1858 and was demolished 1861.

## The Lap Engine and Boulton Coinage.

Matthew Boulton found another sphere for his inventive ability and social responsibility and this was in reforming the coinage.

During the eighteenth century our coinage became quite inadequate for the increasing amount of commerce. Coins were made very labouriously, one at a time, three people operating the clumsy stamping machinery from which the coins 'dribbled out'. As a result, there was a great shortage of money and tradespeople and manufacturers issued their own private money or 'tokens'.

These tokens could be used for exchange in the area round the shop or factory which had issued them but, of course, they were useless further away where people did not know the man who had supplied them and did not want to buy his wares. Boulton had visited foreign mints and gained ideas about improved coining and, when a government committee was set up to consider the problem, he was asked to give his advice. As a result he was then asked to make sets of coins for the government and he built Soho Mint on the site of his manufactory. The machinery which he installed was more accurate and speedy than any which had previously been used. He devised new mechanical means for feeding materials into the machines and arranged the workshops so that men could work in them easily and quickly.

In 1797 the Soho Mint made coins of one penny and two pennies, which were made from copper. The price of copper in 1797 equated to the face value of these

## OPERATING INSTRUCTIONS AND POWER CALCULATION

*Figure 77. Over 100 years ago 'Boulton Pennies' were added to malleable iron in the Birmingham district, when 'specially good metal was required'. 'Boulton Pennies' contained 99 per cent copper.*

*Figure 78. Boulton coins made at the Soho Mint in 1797, the coin shown here is a two penny piece.*

## OPERATING INSTRUCTIONS AND POWER CALCULATION

two coins - the two penny piece actually weighed two ounces.

These coins were not popular with the public because they were too heavy. They became known as cartwheel coinage. However all the evidence suggests that the Lap Engine actually made the copper blanks for these coins in 1797, before Boulton's newly developed coining presses completed the process by impressing the image on each face. Some idea of the output of the Soho Mint can be judged from the fact that between 1797 and 1799 1266 tons of copper had been used to make 45,407,440 penny pieces. A study of Figure 77, will reveal why not many of Boulton's cartwheel coins have survived to the present day.

### Information Sources for the Boiler

As there is no material evidence remaining of the original boiler which provided the steam to power the Lap Engine, the details needed to make the boiler for the model were the result of an extensive search. At first all the local reference libraries were checked, without any success. However two drawings of the actual installations were eventually found in the Science Museum in South Kensington, London. These drawings (Figure 47 and 48) proved invaluable because they clearly showed the boiler setting in relation to the engine and also showed, in elevation, the relative height of the boiler top in relation to the main framework of the engine. From these dimensions, the working floor level could

*Figure 79. The workshop area, seen on the miniature recreation of the Lap Engine. The blacksmith's vice holds a workpiece being cut by a Lancashire hacksaw.*

## OPERATING INSTRUCTIONS AND POWER CALCULATION

be established and this fact was of the upmost importance before the model could be seriously considered.

Because the original drawings of 1788 were too faint to copy, a tracing had to be taken with great care so as not to miss any detail, however small. There appears to be some uncertainty as where to position the flywheel, as lightly pencilled on the drawing the flywheel is projecting through the side wall. The most important details, such as the height, the width and length of the boiler were clear. With this information serious work on the model could begin. Figure 48, shows two other very important points: the primary pump had a wooden launder and the plug tree was originally made of wood in 1788. The drawings are the opposite way around to the actual engine house, hence the word 'reverse' on each print.

In 1779, James Watt developed a printing process for reproducing both drawings and his own correspondence. This process is the same as transfer printing, but as this reversed the image, the print which was on very thin paper, had to be read by holding the drawing up to the light and looking from the back.

As there are no drawings whatever in existence of the actual valves and fittings, all the relevant information needed to complete the boiler was obtained from the books, '*A Treatise on the Steam Engine,' 1827* by John Farey, and '*James Watt and the Steam Engine 1926'*, by H W Dickinson and Rhys Jenkins. All the historical details of this engine by James Watt were gathered together over a period of four years (1978-1982), and were used as a reference when the Lap Engine was constructed in miniature form. However, in the process of building the model, so much information came to light that the project did not seem complete, without writing this account as a permanent record of an engine which has played such an important part in the development of mechanised industries.

## CHAPTER 11

# JAMES WATT AND HIS LIFE IN RETIREMENT

**James Watt - the Final Years**

The two engines designed by James Watt were both precursors to the rapid industrial growth in the latter part of the eighteenth century. The Smethwick Engine set the standard upon which all future coal and mineral mine drainage engines were based. It was the first of James Watt's three-valve engines which allowed the combined force of low pressure steam and the created vacuum from his separate condenser to be used for the first time.

This method of working led to the development, particularly in Cornwall, of using high pressure on the top of the piston with a vacuum, formed by separate condensation of steam on the underside. This became so synonymous with Cornish engines that its principle became known as the Cornish cycle. However, Cornish engineers and Watt reverted to supporting the main oscillating beam within the end gable wall of the engine house and the original engine at Smethwick is thought to be the only installation in the eighteenth century to have the complete engine assembly enclosed within one building. So masterful were the principles used in the design of the Lap Engine, that engines were made using James Watt's ideas for more than a hundred years.

*Figure 80. The parallel motion fitted to an engine made in 1884 by James Watt & Co, almost one hundred years after the mechanism was first fitted to the Lap Engine. This is the magnificently decorated engine and interior of the Papplewick Pumping station, Nottingham.*

The one idea which gave Watt the most satisfaction was never superseded in operation, Watt's parallel motion. It is interesting to compare the parallel motion to the piston rod which was fitted to the Papplewick, Nottinghamshire, pumping engine of 1884 with that of the Lap Engine - the method of operation is identical and yet the Papplewick engine was built over a hundred years later!

The use of a centrifugal governor to control the rotary speed of an engine started with the device fitted to the Lap Engine and has become known as the Watt Governor. This method of rotary speed control was never superseded throughout the period of steam engine design. Watt's contribution to steam engine design is immense. His engines were designed on sound principles which eventually led to production engineering methods used throughout the world in the nineteenth century.

He had taken the rural crafts of the early eighteenth century and assembled them in one place to enable the Soho Manufactory to be successful and in 1796, the Soho Foundry became the first organisation in the world to be devoted solely to designing and building of steam engines.

**Life in Retirement, 1800-1819**

From 1795 onwards, Watt appears to have withdrawn gradually from active participation in the steam engine business. When his partnership with Matthew Boulton finally drew to a close in 1800, it was no wrench but a happy and quiet release. A share of the heavy arrears of royalties which had been collected in the preceding year fell to Watt rendering him well off, and he indulged in the pleasant task of looking out for a landed estate. In 1798 he had fixed upon a

*Figure 81. 'Heathfield', James Watt's cherished house at Handsworth Heath Birmingham, which was designed by Samuel Wyatt and built in 1790.*

property with a farmhouse at Doldowlod on the banks of the Wye between Rhayader and Newbridge, Radnorshire.

Here he was accustomed to spending part of the summer but he was too old to root himself afresh and he returned with zest to his comfortable house, 'Heathfield', at Handsworth Heath, Birmingham, to his mechanical creations and to the society of his old friends. Figure 81 shows Watt's house, 'Heathfield' and Figure 82 shows the Garret Workshop where he spent much of his time in retirement. This workshop is now preserved in the Science Museum, London, in the illustration can be seen his experimental method of reproducing sculptures.

*Figure 82. The Garret Workshop where James Watt conducted many of his experiments in his retirement.*

According to Dickinson and Jenkins, in '*James Watt and the Steam Engine*', Watt briefly returned to Scotland and spent the winter of 1805 in Edinburgh where it was noted by one of Watt's acquaintances, Henry Brougham (later Lord Brougham), that Watt was a constant attendant at their Friday club and, in all the members' private circles, and he was the life of them all. Among the many members of this weekly gathering was Sir Walter Scott, the poet and novelist, who wrote the following account of James Watt.

'There were assembled about half a score of our Northern Lights .. Amidst this company stood Mr Watt, the man whose genius discovered the means of multiplying our national resources to a degree perhaps even beyond his own stupendous powers of calculation and combination, bringing the treasures of the abyss to the summit of the earth - giving the feeble arm of man the momentum of an Afrete commanding manufactures to arise, as the rod of the prophet produced water in the desert - affording the means of dispensing with that time and tide which wait for no man, and of sailing without that wind, which defied the commands and threats of Xerxes himself.'

'This potent commander of the elements this abridger of time and space, this magician, whose cloudy machinery has produced a change in the world, the effects of which, extraordinary as they are, perhaps only now beginning to be felt, was not only the most profound man of science, the most successful combiner of powers and calculation of numbers, as adapted to practical purposes, was not only one of the most generally well-informed, but one of the best and kindest of human beings.

'There he stood surrounded by the little band . . of Northern literati, men not less tenacious, generally speaking, of their own fame and their own opinions, than the national regiments are supposed to be jealous of the high character which they have won upon service. Me thinks I yet see and hear what I shall never see or hear again.'

'In his eighty-fifth year [sic] - a slight slip, or shall we say literary licence? When these words were written Watt was only in his seventy-ninth year, the alert, kind, benevolent old man had his attention alive to everyone's question, his information at everyone's, command.

His talents and fancy overflowed on every subject. One gentleman was a deep philologist he talked with him on the origin of the alphabet as if he had been coeval with Cadmus; another, a celebrated critic - you would have said the old man had studied political economy and *belles-lettres* all his life, - of science it is unnecessary to speak, it was his own distinguished walk'.

'At this time, Watt took pleasure in the society of men younger than himself and this was some consolation for the loss of so many of his older friends as they died. However, he found most satisfaction in the pursuit of new inventions, no longer as a business, but as a hobby in his old age. He occupied himself from 1804 onwards with machines used for copying irregular objects such as busts. No doubt when in France he had seen the 'tour d' portrait' which was a machine, capable of copying, at the same size or at a reduced scale, an object in slight relief -such as a medallion - and this may have given a direction to his labours and stimulated his inventive powers'.

The principle of the machine was to have a rapidly revolving cutting tool guided by a linkage of bars from a feeler, passing over the original object by successive small advances. Watt first modified the machine by mounting the workpiece and the original sculpture on axes, so that they were capable of a complete revolution in synchronisation with one another. In this way the whole surface of a solid could be machined. The design of the framing connecting the feeler and the cutting tool had to be such that absolute rigidity could be maintained, without any vibration or shake. Several of the frames built up from tubes, as well as a complete machine in wood, were in his workshop at Heathfield. Probably it was in connection with the framing that Watt wrote in a letter: 'I return you my best thanks for the experiments and calculations on the rigidity of tubes that you have been so good as to make for me and which prove satisfactory'.

It should also be mentioned that William Murdock helped James Watt in the practical construction of this machine. Watt never took out any patent for it, although he drafted a specification which was dated 21st September, 1814.

He did quite a lot of good work with the machine, as witness the numerous copies of busts that he made for his friends, as well as those preserved in the workshop itself, 'the production' he writes, 'of a young artist entering upon his eightieth year'. Watt intended to make a more accurate machine entirely in metal and even prepared some of the drawings, but he did not live to carry out his intention.

The workshop contains a wonderful collection of tools and apparatus, some without doubt dating from the Glasgow College days, and we must be grateful to the younger James Watt, by whose care its contents were preserved.

It may be said that Watt's career, considered merely from the standpoint of his eminence as an inventor, was in striking contrast with that of other great contemporary inventors - and there were giants in those days - because, unlike most of them, Watt reaped a pecuniary reward for his labours and, as is the way of the world, honours came to him as his years increased.

He was elected a Fellow of the Royal Society of Edinburgh in 1784, and of the Royal Society of London on 24th November 1785, at the same time as Matthew Boulton.

In 1787 Watt was elected a member of the Batavian Society of Rotterdam while in 1806 the University of Glasgow most appropriately honoured itself by conferring on him the degree of Doctor of Laws, 'honoris causa'. In return, in 1808 he founded the Watt Prize in Natural Philosophy and Chemistry.

He was made a corresponding member of the French Academy in 1808 and in 1814 had the singular honour of being elected one of its eight foreign associates. In Birmingham he supported the Church of England. In a letter describing the Church and King riots there in 1791, Watt states: 'I among others, was pointed out as a Presbyterian, though I was never in a meeting house in Birmingham'. In Cornwall, Watt attended chapel on at least one occasion. James Watt was modest to the extreme of self depreciation. Black wrote of him once, 'Were you to be the first publisher of your discoveries, you would do it in such a cold and modest manner that blockheads would conclude there was nothing in it, and rogues would afterwards, by making a trifling variation, vamp off the greater part of it as their own. Watt was cautious in the extreme and as equally unfitted for high enterprise as for high finance'.

He was unremitting in his application; his patience was great; his fertility of intellect was such that he could not avoid scheming improvements on any invention mentioned to him.

Looking back dispassionately over the space of a century and bearing in mind the advances that have been made in the steam engine since Watt's day, we can affirm still that no one has made a greater individual contribution to its development.

The end came quite peacefully, and James Watt died at 'Heathfield' on 25th August 1819 at the age of aged 84. He had been in his usual health the previous month, in fact he had been to London. Watt was buried beside Matthew Boulton in Handsworth Church, Birmingham, on 2nd September. The Watt Chapel was built subsequently, in his will which was dated 7th July 1819, he left to his wife Ann £1,400 per annum, together with the house 'Heathfield' for life and, to his son, James, the residue of the estate. His wife and son were appointed executors. All of his books, drawings, tools, etc. went with the residue. There are two codicils by which he left a number of small legacies which, with the exception of £150 to charities were to old friends, colleagues and servants.

*Figure 83. Sir Francis Chantrey's marble statue of James Watt in the parish Church of St Mary's Handsworth Birmingham.*

Typical of the man is his request that he might be 'interred in the most private manner without show or parades soon after his decease as may be proper. It can hardly be said that this request was faithfully carried out, since more than £700 of the expenses were for the funeral. The will was proven by the executors on 13th October for upwards of £60,000.

Unlike many other great inventors, since his death Watt has held a high place in the estimation of mankind. Perhaps this is due in some measure to the filial conduct of his son who honoured his father by establishing or contributing to the cost of memorials to him. Typical are the statues by Chantrey in the Watt Chapel of Handsworth Church, and the Watt Memorial in Greenock, and also the Watt Memorials in Westminster Abbey.

The last named memorial was set up by public subscription initiated at a public meeting presided over by the Prime Minister, Lord Liverpool, on 18th June 1824. The inscription upon the plinth is from the pen of Lord Brougham. Lofty as its language is, it has the signal merit that every word rings true.

> NOT TO PERPETUATE A NAME
> WHICH MUST ENDURE WHILE THE PEACEFUL ARTS FLOURISH
> BUT TO SHOW
> THAT MANKIND HAVE LEARNED TO HONOUR THOSE
> WHO BEST DESERVE THEIR GRATITUDE
> THE KING
> HIS MINISTERS AND MANY OF THE NOBLES
> AND COMMONERS OF THE REALM
> RAISED THIS MONUMENT TO
>
> **JAMES WATT**
>
> WHO DIRECTED THE FORCE OF AN ORIGINAL GENIUS
> EARLY EXERCISED IN PHILOSOPHIC RESEARCH
> TO THE IMPROVEMENT OF
> THE STEAM ENGINE
> ENLARGED THE RESOURCES OF HIS COUNTRY
> INCREASED THE POWER OF MAN
> AND ROSE TO AN EMINENT PLACE
> AMONG THE MOST ILLUSTRIOUS FOLLOWERS OF SCIENCE
> AND THE REAL BENEFACTORS OF THE WORLD.
>
> BORN AT GREENOCK 1736
> DIED AT HEATHFIELD IN STAFFORDSHIRE 1819.

Lord Brougham wrote: 'It has ever been reckoned by me one of the chief honours of my life that I was called upon to pen the inscription upon the noble monument thus nobly reared'. The inscription is perhaps the finest one of its kind in the English language, Dean Stanley said of it: 'It is not unworthy of the omnigenous knowledge of him who wrote it or of the powerful intellect and vast discovery which it is intended to describe'.

Many have been the tributes to James Watt as a man, that of Wordsworth deserves a place from its great insight: 'I look upon him considering both the magnitude and the universality of his genius, as perhaps the most extraordinary man that this country ever produced; he never sought display, but was content to work in that quietness and humility both of spirit and outward circumstances in which alone all that is truly great and good was ever done'.

We conclude this brief biography of Watt's last years in his own words, the spirit of which has been ever present with us in this work: 'Preserve the dignity of a philosopher and historian; relate the facts and leave posterity to judge. If I merit it some of my countrymen may say: " *Hoc a Scoto factum fuit*".

To commemorate the life of James Watt, a bronze medallion was struck at the

## JAMES WATT'S STANDARD ENGINE

engines. This was an engine which has been central to this story, which was all started by researching the Lap Engine to enable a one sixteenth scale model to be made.

*Figure 84. The bronze medallion struck at the Soho Manufactory to commemorate the life of James Watt in 1819. The engine shown has similar proportions to that of the Lap Engine.*

### The Manufacture of Steam Engines after 1800

The extended patent first taken out by James Watt in 1769 for his separate condenser expired in 1800. Because this patent lasted for a total of thirty-one years, other engine manufacturers found it very difficult to compromise their engines as they could not visualize another way around this patent. Boulton and Watt did not even allow other engine manufacturers to make the separate condenser under licence and it appears ironical that such a beneficial invention should hinder the technological progress of the steam engine for such a long time.

Engineers must have eagerly awaited the year 1800 when engine manufacturers throughout the country could copy the separate condenser and at last build engines as efficient as those which were constructed at Boulton and Watt's Soho Manufactory. A Manchester company even called themselves the Soho Foundry. This company, founded in 1804, had no known connection with Boulton and Watt, but nevertheless produced almost exact copies of the Birmingham engines. A close study of the engraving, Figure 85, will reveal components almost identical to those manufactured in Birmingham, such as a

*Figure 85. The Soho Foundry of Peel and Williams of Manchester, c1804.*

flywheel, an eduction system and some nozzle housings. The cylinder in the right-hand corner of this engraving even has the same number of bolt holes as the end flange had on the Lap Engine!

However, the *real* advance came about in 1804 when the Cornish engineer, Richard Trevithick, used only high pressure steam in a horizontally positioned double-acting cylinder for the first time. This engine was of a very simple construction without any need for the very complicated mechanisms which had been necessary in order to operate the double-acting atmospheric engines of the eighteenth century. Trevithick's double-acting rotary engine worked at 45-50 pounds per square inch steam pressure.

## JAMES WATT'S STANDARD ENGINE

*Figure 86. A pictorial drawing by the author showing how the engine was assembled at the Soho Manufactory in 1788. Shown in this drawing is the waggon boiler in relation to the actual engine.*

## JAMES WATT'S STANDARD ENGINE

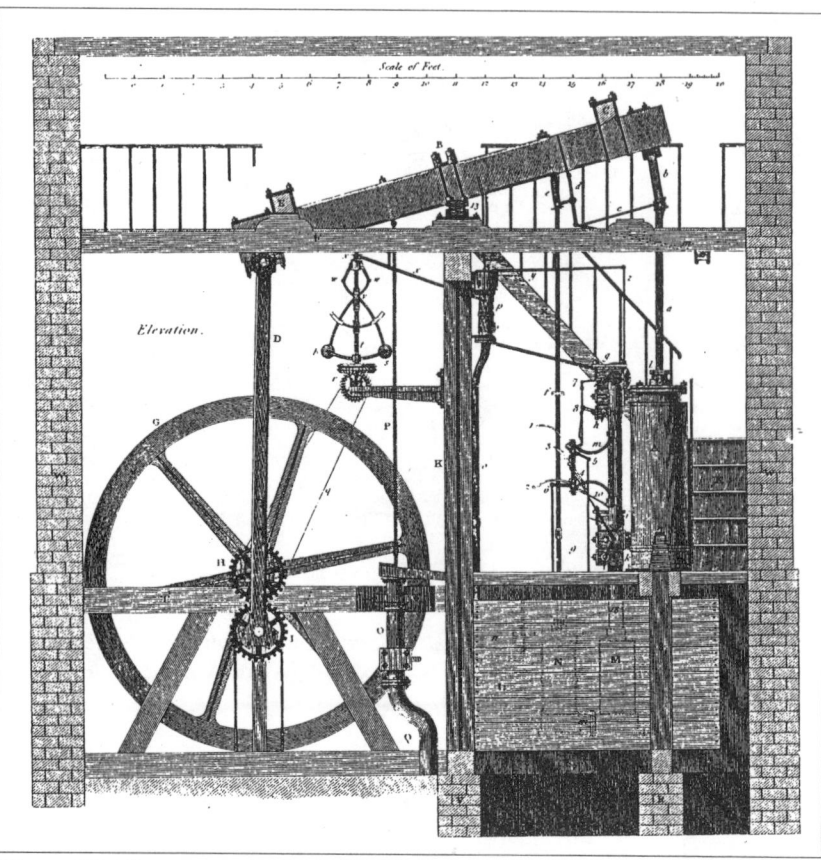

*Figure 87. The engine shown in this drawing became known as the standard ten horse power engine supplied to manufacturers in Great Britain and was exported to many countries throughout the world.*

# CHAPTER 12

# THE SOHO MANUFACTORY

*Figure 88.*
*When the Soho Manufactory opened in 1762, it became the largest factory of its type in the world, and was owned by Matthew Boulton.*
*In the early years the machinery at the Manufactory was driven by water power from Hockley Brook. However when this became inadequate due to increased production, Watt's steam engines were introduced. This is where the Lap Engine was used to drive the polishing machines between the years 1788 and 1858.*
*This is how the manufactory was described in the Birmingham Directory of 1774, ' Four Squares with shops, warehouses etc. for a thousand workmen who, in a great variety of branches, excel in their several departments, not only in the fabrications of Buttons, Boxes, Trinkets etc., in Gold and Silver, but in many other Arts, long predominant in France, which lose their reputation on a comparison with the product of this place'*

## THE SOHO MANUFACTORY

*Figure 89. A photograph of the Soho Manufactory showing how Boulton had transformed a barren heath into an estate with a delightful garden. This picture was taken in the late 1850s.*

*Figure 90. A photograph taken in 1861 shortly before the Manufactory was demolished. It is reproduced from a wet collodion negative.*

## THE SOHO MANUFACTORY

*Figure 91. The Soho Manufactory being demolished in 1861. This was shortly after the death of James Watt Junior. Nothing now remains of this industrial complex which had been conceived by Matthew Boulton in 1762.*

*Figure 92. Finally James Watt's workroom at Heathfield which remained undisturbed until this photograph was taken in 1901. In the foreground can be seen his sculpture copying machine together with completed busts of both Boulton and Watt. The contents of this workshop are now displayed at the Science Museum in South Kensington.*

# BIBLIOGRAPHY

*A Treatise on the Steam Engine* by John Farey, Volume 1 (1827, reprinted by David and Charles 1971)

*James Watt and the Steam Engine.* H. W. Dickinson & R. Jenkins (1927)

*Rees's Manufacturing Industry* (1819-20) Volume Five.

*A Short History of the Steam Engine,* H. W. Dickinson (second edition 1963)

*The Soho Foundry, Birmingham,* W. K. V. Gale, W & T. Avery Ltd. (1946)

Transactions of the Newcomen Society (1961–1998)

*James Watt,* L.T.C. Rolt, B.T. Batsford Ltd (Published 1962)

*Copper in Cast Steel and Iron* Copper Development Association (1937)

*Images of England* Tempus Publishing Limited (1998)

*The Steam Engine of Thomas Newcomen* (L.T.C. Rolt and J.S. Allen) (1977)

# GLOSSARY OF TERMS

**Adze**
A hand tool for cutting away the surface of wood like an axe with an arched blade at right angles to the handle.
**Atmospheric Engine**
The earth's atmospheric pressure is used to power an engine by pressing against a vacuum created on the underside of the piston. .
**Blister Steel**
High carbon steel produced by the cementation process.
**Bore**
A cylindrical hole usually containing a sliding plug or piston.
**Caulking**
A means of closing the seams between wrought iron plates by the use of a blunt chisel.
**Condensate**
Water produced after condensation of the steam inside the cylinder or a separate condenser.
**Condenser**
A sealed vessel in which steam is condensed to create a vacuum.
**Corves**
Basket to put coal in, a man who made them was called a corver.
**Double Acting**
An engine with the piston powered in both directions.
**Expansive Force**
The force produced by steam expanding within a cylinder.
**Factory System**
A system which came into use in the eighteenth century for the manufacture of artefacts which were all produced on one site.
**Flap Valve**
A valve used for the one way passage of water and usually closed by the force of gravity.
**Gib and Cotter**
A metal attachment used to hold two components into a working position.
**Great Lever**
A term used to describe the main oscillating beam of a steam engine.
**Hardware**
Small ware or goods usually made from metal, eg. ironmongery.
**Hogs Heads**
A measure of a volume of water - 53 imperial gallons.
**Indicated Horsepower**
Is calculated by the formula
P x L x A x N divided by 33,000
P = mean pressure on the piston (psi)
L = length of the piston stroke in feet.
A= area of the piston in square inches.
N = number of strokes per minute.
33,000 = work done in foot pounds per minute and is equal to one horse power.
**Injection Valve**
A valve used to admit a controlled volume of water into the condenser.
**Iron Cement**
A compound of iron filings moistened with sal ammoniac which was used to seal the joints on cast iron pipes.
**Mucksand**
A material which is used to make cores used in sand casting - main ingredient horse manure.
**Natural Philosophy**
The forerunner of modem physics.

# GLOSSARY OF TERMS

**Normal Horsepower**
Is an obsolescent term once used to rate the power of an engine.
**Nozzle**
A metal housing which is placed at each end of the powering cylinder containing the steam valves.
**Oakum**
A fibre obtained by the untwisting of old rope used for sealing joints to stop leaking.
**Patent**
A licence which grants the patentee (inventor) the sole right to make or sell his invention.
**Pickle Pot Condenser**
A condenser used on the later Newcomen engines.
**Piston**
A sealed metal plunger which slides in the bore of a cylinder.
**Plug Tree**
A vertical rod attached to the main beam of an engine, used to operate the valve gear positioned below.
**Plummer Block**
A block of metal usually cast iron which is held into position by bolts and also containing a bearing.
**Power of Engines**
Long hundredweights: Thomas Newcomen's method of calculating an engine's power output by, using the pressure of the earth's atmosphere. James Watt calculated the power of his engines by relating them to how much work a standard horse could do in one minute.
**Preventer**
An attachment bolted onto the main beam of an engine used to prevent damage in any emergency, by halting the movement of the oscillating beam.
**Primary Pump**
A pump which is used to draw the water needed to run an engine from a well or stream.
**psi**
Pounds per square inch, usually used to calculate the force exerted by a piston.
**Receiving Tank**
A tank used to contain the water from the air pump.
**Secondary or Hot Water Pump**
A pump which is used to raise the water from the receiving tank into the header tank of the engine.
**Single Acting**
An engine where the piston is powered in one direction only, usually by a vacuum on the underside.
**Snifting Valve**
A valve designed to retain a vacuum, and allow the incoming steam to completely fill the cylinder.
**Stuffing Box**
A sealing box packed with hemp used to seal a circular sliding rod, usually the piston rod.
**Tappets**
Adjustable attachments on the side of the plug tree which were used to operate the levers controlling the flow of steam to the powering cylinder.
**Vacuum**
A space which has had the air removed, or at a very low pressure.
**Vacuum Gauge**
An instrument for checking the pressure inside the condenser, which is calibrated in inches of mercury.

# INDEX

## A
Aeolipyle .................................................................................. 52
Albion Mill ............................................................................. 104
Ammonia ................................................................................. 28
Annual Payment ...................................................................... 55
Arch Head chains .................................................................... 45
Archimedes Principle ............................................................. 70
Atmospheric Engine .......................................................... 12,22
Avery A&T ............................................................................ 120

## B
Bakewell .................................................................................. 56
Beam Calculation .................................................................... 94
Beam ....................................................................................... 44
Beams Spring .......................................................................... 93
Bearing Blocks ................................................................... 48,94
Birmingham Canal Navigation Company ............................... 45
Bob Tumbling ..................................................................... 41,42
Boiler Plates ............................................................................ 27
Boiler Safety Valve ................................................................. 31
Boiler Size ............................................................................... 63
Boiler Making ......................................................................... 65
Boiler Care ............................................................................ 113
Boiler Feed .............................................................................. 69
Boiler Haystack ................................................... 24,25,62,122
Boulton, Matthew ............................................................ 1,14,15
Boulton, Matthew Piers Robinson .......................................... 58
Boulton, Coinage .......................................................... 120,121
Burton-upon-Trent ............................................................ 28,67
Burton Wharf .......................................................................... 67

## C
Calculations Pump Sizes ........................................................ 80
Caulking .................................................................................. 68
Chains ..................................................................................... 45
Cheddleton Flint Mill ......................................................... 50,62
Circular Motion ....................................................................... 10
Coal Consumption ............................................................... 7,56
Coinage, Boulton ........................................................... 120,121
Condensate ............................................................................. 69
Condensation System ............................................................. 75
Condenser Water Cooled ................................................... 55,78
Connecting Rod ................................................................. 49,96
Cooling Tank ........................................................................... 79
Cornwall ..................................................... 14,15,18,55,94,118

# INDEX

Crank and Flywheel ..................................................................48,50
Crank ..................................................................................................14
Cylinder Double Acting................................................................. 99
Cylinder .....................................................................................33,84

## D
Damper Flue....................................................................................... 73
Dickinson and Jenkins ................................................................57,64,89
Double Acting Cylinder.................................................................. 99
Double Acting Engine ..................................................................18
Drawing of 13$^{th}$ December 1788............................................................. 106
Drawings Dated 29$^{th}$ July 1788........................................................ 63,123
Dudley Castle Engine......................................................................... 1,22

## E
Eduction Pipe.....................................................................................71
Engine Speed..................................................................................... 118
Engines made after 1800 ..................................................................131
English Oak..................................................................................... 44,109
Epicyclic Gears.................................................................................. 99
Exhibition, Midlands Model Engineering ...............................................60,62

## F
Farey, John ........................................................................20,50,52,79
Flywheel .......................................................................6,8,10,13,19,50,51,102
Furnace............................................................................................... 24

## G
Grate Sizes........................................................................................ 64
Gauge Pressure................................................................................. 71
Gear Ratio .......................................................................................100
Gib and Cotter .................................................................................102
Gear Boxed ......................................................................................102
Governor ....................................................................................104,107
Garret Workshop............................................................................ 128

## H
Harmonic Motion ...........................................................................121
Haystack Boiler .........................................................................25,60
'Heathfield'......................................................................................126
Hockley Brook ...........................................................................81,115
Hornblower ......................................................................................55
Hulls, Jonathan................................................................................... 6

# INDEX

## I
Inspection Hatch ...... 24
Injection Valve ...... 78
Iron Cement ...... 118

## J
Japanese Oak ...... 44
Jointing Material ...... 113

## L
Lap Engine Drawing ...... 133
Lap Engine ...... 1,57,59.61
Lathe ...... 6,35
Law ...... 52
Lead Mine ...... 31
Lloyd, Charles ...... 27,28,67
Lubrication ...... 84,112

## M
Maintenance ...... 112

Mead, Thomas ...... 104,105
Mechanism, Unusual ...... 70
Midlands Model Engineering Exhibition ...... 60,62
Miner's Friend ...... 5
Miniature Cylinder ...... 35
Mucksand ...... 33
Murdock, William ...... 99
Museum, South Kensington ...... 1,57,60,85
Museum, Patent ...... 58

## N
Newcomen Society ...... 21
Newcomen, Thomas ...... 11,14,18,42,55,89
Nozzle or Valve ...... 85,86

## O
Oakum ...... 69

## P
Papin, Denis ...... 4,8
Parallel Motion Component Parts ...... 91
Parallel Motion ...... 89,90
Patent ...... 13,14,16,52,53,75,102,105

# INDEX

Pickard, James ...................................................................... 1,13,99
Piston .................................................................................... 36
Piston Seal ............................................................................ 37
Power Measurement ........................................................... 119
Preventer .............................................................................. 95
Process Machinery ........................................................... 13,19
Pump ............................................................................... 79,80

## R
Rack ..................................................................................... 10
Ratchet and Pawl ................................................................. 13
Reeves, A.J .......................................................................... 66
Rod Connecting ................................................................... 96
Rotary Motion Without Crank ............................................ 99
Rotary Motion ..................................................................... 10

## S
Safety Valve ......................................................................... 68
Savery, Thomas ..................................................................... 5
Schematic Drawing ........................................................... 116
Seal ...................................................................................... 37
Sealing Steam Pipes .......................................................... 118
Single Acting ....................................................................... 46
Smeaton, John ........................................................ 11,12,81,82,83
Smethwick Engine ...................................................... 1,44,45,82
Snow Hill ...................................................................... 1,13,14,19
Soho Foundry .................................................................... 120
Soho Manufactory ....................................................... 1,16,57
Soho Mint .......................................................................... 122
Southern, John ............................................................. 19,44,48
Spear .................................................................................... 96
Speed Increase ..................................................................... 96
Speed Control .................................................................... 106
Spray Water ......................................................................... 40
Standard Engine ................................................................ 129
Starting the Engine ....................................................... 39,117
Stewart, John .................................................................. 10,11
Stone or Float ...................................................................... 70
Stuffing Boxes ................................................................. 69,84
Sun and Planet ........................................................... 53,99,101
Sweden ................................................................................ 66

## T
Trevithick, Richard ............................................................. 75

## INDEX

Tribute to James Watt ............................................................126,127,128
Trundle Wheels ..................................................................................10

## V
Vacuum.............................................................................................. 38
Valves.......................................................................................39,68,71
Valve Operation................................................................................. 87
Valve Gear .......................................................................................110

## W
Wasborough, Matthew................................................................ 1,13,99
Water Level Measurement ..................................................................72
Water Level.............................................................................. 30,117
Water Control .....................................................................................69
Waterwheel...............................................................................10,50,83
Watt, James .................................................. 1,14,15,55,124,125,126,127,128
Westminster Abbey ....................................................................129,130
Wheal Busy ........................................................................................55
Wheal Fortune ....................................................................................55
Wilkinson, John........................................................ 16,34,35,78,79,83,118
Wooden Frame of Engine................................................................ 109
Wrought Iron ......................................................................................66

## X
Xerxes ..............................................................................................127